Quantum Chemistry

A concise introduction for students of physics, chemistry,
biochemistry and materials science

T0292399

Quantum Chemistry

A concise introduction for students of physics, chemistry, biochemistry and materials science

Ajit J Thakkar

Department of Chemistry, University of New Brunswick, Fredericton, Canada

Morgan & Claypool Publishers

Rights & Permissions
To obtain permission to re-use copyrighted material from Morgan & Claypool Publishers, please contact info@morganclaypool.com.

ISBN   978-1-627-05416-4 (ebook)
ISBN   978-1-627-05417-1 (print)

DOI   10.1088/978-1-627-05416-4

Version: 20140601

IOP Concise Physics
ISSN 2053-2571 (online)
ISSN 2054-7307 (print)

A Morgan & Claypool publication as part of IOP Concise Physics

Morgan & Claypool Publishers, 40 Oak Drive, San Rafael, CA, 94903, USA

Contents

Preface   **viii**

1   **Molecular symmetry**   1-1

1.1   Symmetry operations and elements   1-1

     1.1.1   Rotations around axes   1-1

     1.1.2   Reflections through symmetry planes   1-4

     1.1.3   Inversion through a center of symmetry   1-5

     1.1.4   Improper rotations around improper axes   1-6

1.2   Classification of molecular symmetry   1-6

1.3   Implications of symmetry   1-9

     Problems   1-10

2   **Basic quantum mechanics**   2-1

2.1   Wave functions specify a system's state   2-1

2.2   Operators represent observables   2-2

     2.2.1   Operators   2-2

     2.2.2   Quantum chemical operators   2-4

2.3   Schrödinger's equation   2-5

2.4   Measured and average values   2-6

     Problems   2-7

3   **Translation and vibration**   3-1

3.1   A particle in a wire   3-1

     3.1.1   Solving the Schrödinger equation   3-2

     3.1.2   The energies are quantized   3-3

     3.1.3   Understanding and using the wave functions   3-4

3.2   A harmonic oscillator   3-5

     3.2.1   Molecular vibrations   3-7

     Problems   3-8

4   **Symmetry and degeneracy**   4-1

4.1   A particle in a rectangular plate   4-1

4.2   Symmetry leads to degeneracy   4-2

4.3   Probabilities in degenerate states   4-4

4.4	Are degenerate wave functions unique?	4-6
4.5	Symmetry of wave functions	4-7
	Problems	4-8

5	**Rotational motion**	5-1
5.1	A particle on a ring	5-1
5.2	A particle on a sphere	5-4
	5.2.1 Rotational wave functions	5-6
5.3	The rigid rotor model	5-7
	Problems	5-8

6	**The hydrogen atom**	6-1
6.1	The Born–Oppenheimer approximation	6-1
6.2	The electronic Hamiltonian	6-2
6.3	The hydrogen atom	6-3
	6.3.1 Energy levels	6-4
	6.3.2 Orbitals	6-5
	6.3.3 Electron density and orbital size	6-6
	6.3.4 Spin angular momentum	6-8
6.4	Hydrogen-like ions	6-9
	Problems	6-9

7	**A one-electron molecule: H_2^+**	7-1
7.1	The LCAO model	7-2
7.2	LCAO potential energy curves	7-3
7.3	The variation method	7-5
7.4	Beyond the LCAO model	7-6
7.5	Force constant and dissociation energy	7-7
7.6	Excited states	7-9
	Problems	7-10

8	**Many-electron systems**	8-1
8.1	The helium atom	8-1
8.2	Spin and the Pauli postulate	8-2
8.3	Electron densities	8-4
8.4	The Hartree–Fock model	8-4
	8.4.1 Matrix formulation	8-6

8.5   Atoms   8-7
8.6   Diatomic molecules   8-9
8.7   The Kohn–Sham model   8-11
      Problems   8-13

9   Qualitative MO theory   **9-1**
9.1   The Hückel model   9-1
9.2   Cumulenes   9-3
9.3   Annulenes   9-4
9.4   Other planar conjugated hydrocarbons   9-6
9.5   Charges, bond orders, and reactivity   9-7
9.6   The Hückel model is not quantitative   9-9
      Problems   9-10

10   Computational chemistry   **10-1**
10.1 Computations are now routine   10-1
10.2 So many choices to be made   10-2
      10.2.1 Selection of a basis set   10-2
      10.2.2 Selecting a functional   10-4
      10.2.3 Heavy atoms and relativistic effects   10-4
      10.2.4 Accounting for a solvent   10-5
10.3 Practical calculations   10-5
      Further study   10-7

Appendices

A   Reference material   **A-1**
      Matrices and determinants   A-1
      Miscellaneous   A-2
      Table of integrals   A-3
      Conversion factors   A-4
      Constants and Greek letters   A-4
      Equation list   A-5

B   Problem hints and solutions   **B-1**

Preface

All chemists and many biochemists, materials scientists, engineers, and physicists routinely use spectroscopic measurements and electronic structure computations to assist and guide their work. This book is designed to help the non-specialist user of these tools achieve a basic understanding of the underlying concepts of quantum chemistry. The emphasis is on explaining ideas rather than on the enumeration of facts and/or the presentation of procedural details. The book can be used to teach introductory quantum chemistry to second-or third-year undergraduates either as a stand-alone one-semester course or as part of a physical chemistry or materials science course. Researchers in related fields can use the book as a quick introduction or refresher.

The foundation is laid in the first two chapters which deal with molecular symmetry and the postulates of quantum mechanics, respectively. Symmetry is woven through the narrative of the next three chapters dealing with simple models of translational, rotational, and vibrational motion that underlie molecular spectroscopy and statistical thermodynamics. The next two chapters deal with the electronic structure of the hydrogen atom and hydrogen molecule ion, respectively. Having been armed with a basic knowledge of these prototypical systems, the reader is ready to learn, in the next chapter, the fundamental ideas used to deal with the complexities of many-electron atoms and molecules. These somewhat abstract ideas are illustrated with the venerable Hückel model of planar hydrocarbons in the penultimate chapter. The book concludes with an explanation of the bare minimum of technical choices that must be made to do meaningful electronic structure computations using quantum chemistry software packages.

I urge readers who may be afraid of tackling quantum chemistry to relax. Rumors about its mathematical content and difficulty are highly exaggerated. Comfort with introductory calculus helps but an open mind and some effort are much more important. You too can acquire a working knowledge of applied quantum chemistry just like the vast majority of students who have studied it. Some tips for studying the material are listed below.

1. The material in later chapters depends on earlier ones. There are extensive back references throughout to help you see the connections.
2. Solving problems helps you learn. Make a serious attempt to do the end-of-chapter problems before you look at the solutions.
3. You do need to learn basic facts and terminology in addition to the ideas.
4. Study small amounts frequently. Complex ideas take time to sink in.

This book grew from the quantum chemistry course that I have taught at the University of New Brunswick since 1985. During the first few years of teaching it, I was unable to find a text book that treated all the topics which I taught in a way I liked. So in the fall of 1994, I wrote a set of 'bare bones' notes after each lecture and distributed them during the next one. The encouraging and positive response of the students kept me going to the end of the course. Having arrived at a first draft in this manner, the bare bones were expanded over the next few years. Since then, this book has been revised over and over again using both explicit and implicit feedback from students who have taken my course; it has been designed with their verbal and non-verbal responses to my lectures, questions, problems, and tests in mind.

Fredericton   Ajit J Thakkar
31 March 2014

Quantum Chemistry

A concise introduction for students of physics, chemistry, biochemistry and materials science

Ajit J Thakkar

Chapter 1

Molecular symmetry

1.1 Symmetry operations and elements

Symmetry is all around us. Most people find symmetry aesthetically pleasing. Molecular symmetry imposes constraints on molecular properties[1]. A *symmetry operation* is an action that leaves an object looking the same after it has been carried out. A *symmetry element* is a point, straight line, or plane (flat surface) with respect to which a symmetry operation is carried out. The center of mass must remain unmoved by any symmetry operation and therefore lies on all symmetry elements. When discussing molecular symmetry, we normally use a Cartesian coordinate system with the origin at the center of mass. There are five types of symmetry operation. The identity operation E does nothing and is included only to make a connection between symmetry operations and *group theory*. The other four symmetry operations—rotations C_n, reflections σ, inversion i, and improper rotations S_n—are described next.

1.1.1 Rotations around axes

A symmetry axis C_n, of order n, is a straight line about which $(1/n)$th of a full 'turn' (a rotation by an angle of $360°/n$) brings a molecule into a configuration indistinguishable from the original one. A C_n axis must pass through the center of mass. A C_1 axis corresponds to a $360°$ rotation and so it is the same as the identity operation: $C_1 = E$. A C_2 axis has a $360°/2 = 180°$ rotation associated with it. In H_2O,

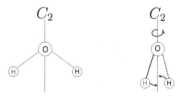

---

[1] As Eugene Wigner said, symmetry provides 'a structure and coherence to the laws of nature just as the laws of nature provide a structure and coherence to a set of events'.

the line bisecting the HOH angle is a C_2 axis; rotation about this axis by 180° just interchanges the two hydrogen nuclei. If the z axis is a C_2 axis, then its action on a nucleus is to move it from its original position (x, y, z) to $(-x, -y, z)$. Thus

$$C_2(z) \begin{bmatrix} x \\ y \\ z \end{bmatrix} = \begin{bmatrix} -x \\ -y \\ z \end{bmatrix}. \tag{1.1}$$

A C_2 axis generates only one unique symmetry operation because two 180° rotations bring an object back to its original configuration; that is, $C_2C_2 = C_2^2 = E$. Each of the objects A, B and C in figure 1.1 has exactly one C_2 axis. The C_2 axis is along the x axis in object A, along the y axis in object B, and along the z axis in object C. The more symmetrical object D in figure 1.1 has three C_2 axes, one along each of the x, y and z axes.

A square has a C_4 axis of symmetry as illustrated in figure 1.2. Performing two successive C_4 or 360°/4 = 90° rotations has the same effect as a single C_2 or 180° rotation; in symbols, $C_4^2 = C_2$. Hence for every C_4 axis there is always a collinear C_2 axis. Moreover, $C_4^4 = C_2^2 = E$ and so a C_4 axis generates only two unique symmetry operations, C_4 and C_4^3. A clockwise C_4^3 rotation is the same as a counter clockwise

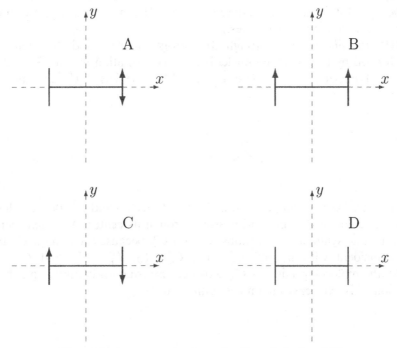

Figure 1.1. Can you see the C_2 axes in objects A, B, C, and D?

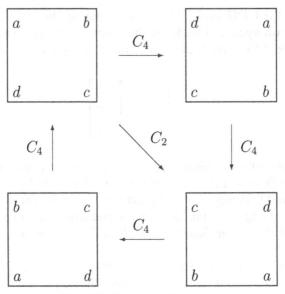

Figure 1.2. C_4 symmetry in a square.

C_4 rotation. We adopt the convention that all rotations are clockwise. A C_4 axis can be found, for example, along each S–F bond in the octahedral molecule SF_6, and along the axial I–F bond in the square pyramidal IF_5 molecule. (Tip: nuclei not on C_n occur in sets of n equivalent ones.)

In NH_3, the line passing through the nitrogen nucleus and the center of the triangle formed by the hydrogen nuclei is a C_3 axis; rotation by $360°/3 = 120°$ permutes the H nuclei ($a \to b$, $b \to c$, $c \to a$). Methane has a C_3 axis along each C–H bond.

A C_3 axis generates two unique symmetry operations, C_3 and C_3^2. Benzene has a C_6 axis perpendicular to the ring and passing through its center. A C_6 axis generates only two unique symmetry operations, C_6 and C_6^5, because a C_3 and a C_2 axis are always coincident with it, and $C_6^2 = C_3$, $C_6^3 = C_2$, $C_6^4 = C_3^2$, and $C_6^6 = E$. In O=C=O, the molecular axis is a C_∞ axis because rotation by any angle, however small, about this axis leaves the nuclei unmoved.

The z axis is taken along the *principal symmetry axis* which is defined as the C_n axis with the highest order n. For example, the C_6 axis is the principal axis in benzene. If there are several C_n axes of the highest n, then the principal axis is the one passing through the most nuclei. For example, ethene (C_2H_4) has three C_2 axes and the principal axis is the one passing through both carbons. A planar molecule that has its principal axis in the molecular plane, like ethene but unlike benzene, is placed in the yz plane.

$$
\begin{array}{c}
H \\[-2pt] \diagdown \\
\end{array}
C=C
\begin{array}{c}
\diagup H \\
\end{array}
$$

1.1.2 Reflections through symmetry planes

A plane is a symmetry plane σ if reflection of all nuclei through this plane sends the molecule into an indistinguishable configuration. A symmetry plane contains the center of mass and bisects a molecule. If the symmetry plane is the xy plane, then its action on a nucleus is to move it from its original position (x, y, z) to $(x, y, -z)$. Thus,

$$
\sigma_{xy} \begin{bmatrix} x \\ y \\ z \end{bmatrix} = \begin{bmatrix} x \\ y \\ -z \end{bmatrix}. \tag{1.2}
$$

A symmetry plane generates only one unique symmetry operation because reflecting through it twice brings a molecule back to its original configuration. Hence $\sigma^2 = E$. A symmetry plane is also called a mirror plane.

The xy plane is a symmetry plane for each of the planar objects A, B, C and D in figure 1.1. Objects A and D also have the xz plane as a plane of symmetry. Objects B and D have a yz symmetry plane. Object C has no other planes of symmetry. Thus object D has three planes of symmetry, objects A and B each have two but object C has only one plane of symmetry. Any planar molecule, such as benzene, has its molecular plane as a plane of symmetry because reflection across the molecular plane leaves all nuclei unmoved.

NH_3 has three planes of symmetry each of which contains an N–H bond and is perpendicular to the plane containing the three hydrogen atoms. (Tip: nuclei not on σ occur in equivalent pairs.) The three symmetry planes are geometrically equivalent, and the corresponding reflections are said to form a *class*. Operations in the same class can be converted into one another by application of some symmetry operation of the group or equivalently by a suitable rotation of the coordinate system. The identity operation always forms a class of its own.

A symmetry plane perpendicular to the principal symmetry axis is called a *horizontal* symmetry plane σ_h. Symmetry planes that contain the principal symmetry axis are called *vertical* symmetry planes σ_v. A vertical symmetry plane that bisects the angle between two C_2 axes is called a *dihedral* plane σ_d. The distinction between σ_v and σ_d planes is unimportant, at least in this book. For example, H_2O has two vertical symmetry planes: the molecular plane and one perpendicular to it. The intersection of the two planes coincides with the C_2 axis. The molecular plane of a planar molecule can be either horizontal as in C_6H_6 or vertical as in H_2O. Benzene also has six symmetry planes perpendicular to the ring and containing the C_6 axis. These six planes separate into two classes: three containing CH bonds and three containing no nuclei. One class of planes is called vertical and the other dihedral; in this example, either class could be called vertical. Linear molecules like HCl and HCN have an infinite number of vertical symmetry planes. Some of them, such as N_2 and CO_2, have a horizontal symmetry plane as well. Reflection in σ_h is always in a class by itself.

1.1.3 Inversion through a center of symmetry

If an equivalent nucleus is reached whenever a straight line from *any* nucleus to the center of mass is continued an equal distance in the opposite direction, then the center of mass is also a center of symmetry. Since the center of mass is at the coordinate origin $(0, 0, 0)$, the inversion operation i moves an object from its original position (x, y, z) to $(-x, -y, -z)$. Thus

$$ i \begin{bmatrix} x \\ y \\ z \end{bmatrix} = \begin{bmatrix} -x \\ -y \\ -z \end{bmatrix}. \tag{1.3} $$

Objects C and D in figure 1.1 have a center of symmetry but objects A and B do not. C_6H_6, SF_6, C_2H_4 and the object in figure 1.3 have a center of symmetry but CH_4

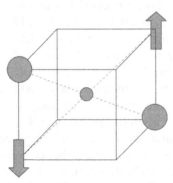

Figure 1.3. This object has a center of symmetry. There would be no center of symmetry if the arrows pointed in the same direction.

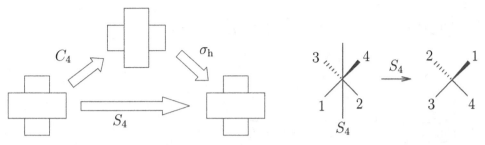

Figure 1.4. A non-trivial S_4 axis with no coincident C_4 axis is shown for stacked erasers on the left and for methane on the right.

does not. (Tip: nuclei not on i occur in equivalent pairs.) Inversion generates only one unique symmetry operation because $i^2 = E$. Inversion forms a class by itself.

1.1.4 Improper rotations around improper axes

An improper rotation S_n is a rotation by $360°/n$ about an axis followed by a reflection in a plane perpendicular to the axis. Thus $S_n = \sigma C_n$. No unique symmetry operations are generated by an S_1 or S_2. Note that $S_1 = \sigma C_1 = \sigma$ because $C_1 = E$, and that $S_2 = \sigma C_2 = i$ as can be seen by combining equations (1.1)–(1.3). Hence only S_n with $n \geqslant 3$ are normally called S_n. Molecules which have both a C_n and a σ_h must have an S_n. For example, benzene has an S_6 axis coincident with its C_6 axis because it has a σ_h.

However, an object or molecule need not have a σ or a C_n to have an S_n. Non-trivial S_4 axes are illustrated in figure 1.4 for a crossed stack of erasers and for methane. Figure 1.4 shows that methane has three S_4 axes, each of which bisects two HCH angles, even though it has neither a C_4 axis nor any symmetry planes perpendicular to an S_4 axis. An S_4 axis generates only two unique symmetry operations, S_4 and S_4^3, because $S_4^2 = \sigma C_4 \sigma C_4 = \sigma^2 C_4^2 = E C_2 = C_2$ and $S_4^4 = S_4^2 S_4^2 = C_2^2 = E$.

1.2 Classification of molecular symmetry

Objects cannot have an arbitrary collection of symmetry elements. For example, it is impossible to have a molecule in which there is a C_3 axis and only one σ_v. A rotation by 120° about the C_3 axis carries the σ_v into a different plane, say P. Since C_3 is a symmetry axis, this new configuration of the molecule must be indistinguishable from the original one. However, for this to be so, the plane P must be a σ_v plane as well. Clearly, a C_3 axis and one σ_v plane imply the existence of two more σ_v planes.

Mathematicians have worked out all possible groups of symmetry operations. Their results can be used to classify molecules by symmetry. Since all the symmetry elements of a molecule must intersect in at least one point, the symmetry groups are

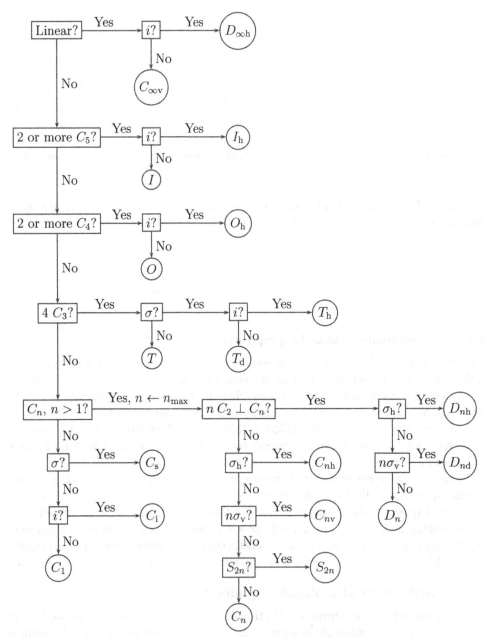

Figure 1.5. Flow chart for determining point group symmetry.

called *point groups*. Each group is designated by a symbol called the *Schoenflies symbol*.

An atom has spherical symmetry and belongs to the K point group. To assign a molecule to a point group, use the flow chart given in figure 1.5. The first step is to decide whether the molecule is linear (all atoms on a straight line). If it is linear, then it

Table 1.1. Characteristic symmetry elements of point groups. $n \geqslant 2$.

Simple		Single-axis groups		Dihedral groups	
C_1	E	C_n	C_n	D_n	$C_n, nC_2 \, (\perp C_n)$
C_s	σ	C_{nv}	$C_n, n\sigma_v$	D_{nd}	$C_n, nC_2 \, (\perp C_n), n\sigma_d, S_{2n}$
C_i	i	C_{nh}	C_n, σ_h	D_{nh}	$C_n, nC_2 \, (\perp C_n), n\sigma_v, \sigma_h$
		Infinite groups		Platonic groups	
		$C_{\infty v}$	$C_\infty, \infty\sigma_v$	T_d	$4C_3, 3C_2, 6\sigma_d, 3S_4$
		$D_{\infty h}$	$C_\infty, \infty\sigma_v, i, \sigma_h$	O_h	$3C_4, 4C_3, i, 6\sigma_d, 3\sigma_h$
		K	$\infty \, C_\infty$	I_h	$6C_5, 10C_3, i, 15\sigma$

has $C_{\infty v}$ or $D_{\infty h}$ symmetry depending on whether or not it has an inversion center. For example, carbon dioxide (O=C=O) has $D_{\infty h}$ symmetry but HCN has $C_{\infty v}$ symmetry.

If the molecule is not linear, then search for non-trivial axes of rotation C_n with $n > 1$. It helps to know that if there is a C_n axis, then all the off-axis nuclei can be separated into sets of n equivalent nuclei. If there are multiple C_n with $n > 2$, then the molecule belongs to a high-symmetry 'Platonic' group. Six C_5 axes indicate I_h, the point group of a perfect icosahedron or pentagonal dodecahedron, or the rare I, which has only the pure rotations of an icosahedron. Buckminsterfullerene C_{60} has I_h symmetry. Three C_4 axes indicate O_h, the point group of a cube or a perfect octahedron like SF_6, or the rare O which has only the pure rotations of an octahedron. Four C_3 axes and no C_4 axes indicate T_d, the group of a perfect tetrahedron like methane, or the rare T which has only the rotations of T_d, or T_h obtained by combining an inversion center with the rotations of T. The I, O, T_h, and T point groups are chemically rare.

If there are no C_n axes at all with $n > 1$, the molecule is of low symmetry and belongs to (a) C_s if there is a symmetry plane, (b) C_i if there is a center of inversion, and (c) C_1 otherwise. If there are some C_n with $n > 1$, choose a principal axis with the maximum n. From this point on, n is the fixed number that you determined in this step. Check for nC_2 axes perpendicular to the principal axis of symmetry. Next, search for a horizontal plane of symmetry, σ_h. Don't assume that the molecular plane in a planar molecule is a σ_h. For example, the molecular plane in benzene is a σ_h but the molecular plane in H_2O is a σ_v. On those rare occasions when you have to look for an S_{2n} axis, bear in mind that $2n$ is always even and that $2n \geqslant 4$ because $n > 1$. Molecules with D_n or S_{2n} symmetry are uncommon. For example, $C(C_6H_5)_4$ has S_4 symmetry, and ethane in a conformation that is neither staggered nor eclipsed has D_3 symmetry. Use table 1.1 to check for all the symmetry elements characteristic of the point group.

Practice finding the point groups[2] for the molecules in figure 1.6. Visualization software that allows rotation of a molecule's ball-and-stick image in three dimensions is helpful. Newman projections, as taught in organic chemistry, help you see D_n, D_{nd}, and D_{nh} symmetry.

---

[2] From left to right, first row: C_1, C_i, C_s, C_2; second row: C_{2v}, C_{3v}, C_{4v}, C_{2h}; third row: D_{2h}, D_{3h}, D_{6h}, O_h; fourth row: D_{2d}, D_{3d}, T_d.

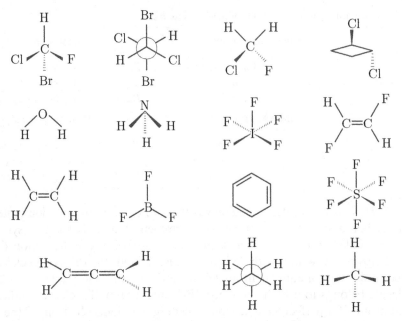

Figure 1.6. Molecules with various symmetries.

1.3 Implications of symmetry

The dipole moment of a molecule should not be changed either in direction or in magnitude by a symmetry operation. This invariance to symmetry operations can be realized only if the dipole moment vector is contained in each of the symmetry elements. For example, an inversion center, more than one C_n axis, and a horizontal symmetry plane all eliminate the possibility of a dipole moment. Therefore, a molecule can have a non-zero dipole moment only if it belongs to one of the point groups C_1, C_s, C_n or C_{nv}. Thus H_2O with C_{2v} symmetry can and does have a non-zero dipole moment, but CO_2 with $D_{\infty h}$ symmetry cannot and does not have a non-zero dipole moment.

A *chiral* molecule is one that cannot be superimposed on its mirror image. Thus, a molecule can be chiral only if it does not have a symmetry element that converts a right-handed object to a left-handed one. In other words, a molecule can be chiral only if it does not have a plane of symmetry or an inversion center or an improper axis of symmetry S_n. Since $S_1 = \sigma$ and $S_2 = i$, we can simply say that the presence of an improper axis of symmetry rules out chirality. A molecule can be chiral only if it belongs to a C_n or D_n point group.

In discussions of rotational spectroscopy, it is usual to classify molecules into four kinds of *rotors* or *tops*. The correspondence between that classification and point groups is simple. *Linear rotors* are $C_{\infty v}$ or $D_{\infty h}$ molecules. *Spherical tops* contain more than one C_n axis with $n \geqslant 3$ as in T_d, O_h or I_h molecules. *Symmetric tops* are molecules that contain one and only one C_n axis with $n \geqslant 3$ or an S_4 axis, and thus belong to C_n, C_{nv}, C_{nh}, D_n, D_{nh} or D_{nd} with $n \geqslant 3$ or D_{2d} or S_n ($n = 4, 6, 8, \ldots$).

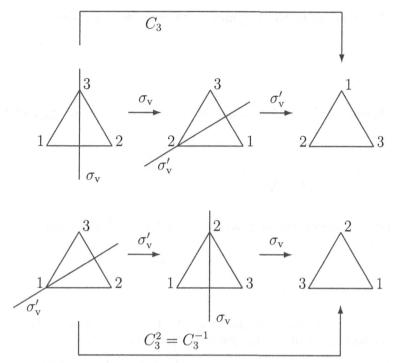

Figure 1.7. Non-commutativity of reflections in an equilateral triangle.

Asymmetric tops are molecules that do not contain any C_n axis with $n \geqslant 3$ or S_4 axis, and thus belong to C_1, C_i, C_s, C_2, C_{2v}, C_{2h}, D_2 or D_{2h}.

Two symmetry operations, $\mathcal{O}_1$ and $\mathcal{O}_2$, are said to commute if the result of carrying out one after the other does not depend upon the order in which they are carried out. That is $\mathcal{O}_1$ and $\mathcal{O}_2$ commute if $\mathcal{O}_1\mathcal{O}_2 = \mathcal{O}_2\mathcal{O}_1$ where $\mathcal{O}_1\mathcal{O}_2$ means first do $\mathcal{O}_2$ and then do $\mathcal{O}_1$. Symmetry operations do not always commute. For example, figure 1.7 shows that, in an equilateral triangle, reflections in the σ_v do not commute with one another; in symbols, we write $\sigma_v'\sigma_v \neq \sigma_v\sigma_v'$. Figure 1.7 also shows that $\sigma_v'\sigma_v = C_3$ and $\sigma_v\sigma_v' = C_3^2$.

Groups in which each symmetry operation commutes with every other symmetry operation are called *Abelian*. Every element of an Abelian group forms a class by itself. Note that the symmetry group of an asymmetric top molecule is always an Abelian point group. The energy levels of molecules with Abelian symmetry have a special simplicity as we shall see in section 4.2.

Problems (see appendix B for hints and solutions)

1.1 Which of the molecules in figure 1.6 has a center of inversion?

1.2 Suppose the z axis is a C_4 axis of symmetry. What will be the coordinates of a nucleus after a clockwise C_4 rotation if its coordinates were (x, y, z) before the rotation?

1.3 Use sketches to show all the symmetry elements in naphthalene.

1.4 Use sketches to show all the symmetry elements in the following molecules:

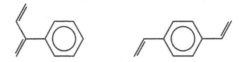

1.5 Find two planes of symmetry and three C_2 axes in allene (C_3H_4).

$$\underset{H}{\overset{H}{\diagdown}}C=C=C\underset{H}{\overset{H}{\diagup}}$$

Use sketches to show the symmetry elements. Drawing Newman diagrams (projections) in the manner of organic chemistry books is helpful.

1.6 Find the symmetry point group for each of the following molecules. Which molecules are polar and which are chiral?

1.7 Find the symmetry point group for each of the following molecules. Which molecules are polar and which are chiral?

1.8 A molecule has three C_2 axes that are perpendicular to each other, and no other non-trivial symmetry elements. Can such a molecule have a non-zero dipole moment? Can it be chiral? Explain without reference to the point group of the molecule.

IOP Concise Physics

Quantum Chemistry
A concise introduction for students of physics, chemistry, biochemistry and materials science
Ajit J Thakkar

Chapter 2

Basic quantum mechanics

2.1 Wave functions specify a system's state

Newton's laws do not describe correctly the behavior of electrons in molecules; instead, quantum mechanics is required. Following early foundation work by Max Planck, Albert Einstein, Niels Bohr, and Louis de Broglie, modern quantum mechanics was discovered in the 1920s, primarily by Werner Heisenberg, Max Born, Erwin Schrödinger, Paul Dirac, and Wolfgang Pauli. All nine won Physics Nobel Prizes.

The justification for quantum mechanics is that it provides an accurate description of nature. As Bohr said, 'It is wrong to think that the task of science is to find out how Nature *is*. Science concerns what *we can say* about Nature.'

In quantum mechanics, the *state* of a system is completely specified by a function called the *time-dependent* state or *wave function*. It is a function of the time *t*, and of the three position coordinates of each of the particles in the system. A system which is not subject to time-varying external forces can be described by a *time-independent* wave function ψ (read ψ as sigh), which is a function of the position coordinates but does not depend on the time. In this book we focus exclusively on such systems. Hence, all wave functions will be time-independent unless explicitly stated otherwise.

For example, the wave function of a one-particle system can be written as $\psi(x, y, z)$ where (x, y, z) are the Cartesian coordinates of the position of the particle. The wave function contains all the information that can be known about the system. Max Born's interpretation of the wave function is that $|\psi|^2$ is a *probability density*. Thus, for a one-particle system, the probability that the particle is in a tiny box centered at (x, y, z) with sides dx, dy, and dz is given by $|\psi(x,y,z)|^2 \, dx \, dy \, dz$, where $dx \, dy \, dz$ is called the volume element[1].

---

[1] The probability density $|\psi|^2$ is real and non-negative, as it must be, even if the wave function ψ is complex-valued. In that case, $|\psi|^2 = \psi^*\psi$ where the asterisk denotes complex conjugation; appendix A has a brief review of complex numbers. In this book, we will work only with real-valued wave functions, and so $|\psi|^2$ will be the same as ψ^2.

To ensure that the sum of the probabilities for finding the particle at all possible locations is finite, the wave function ψ must be *square-integrable*: $\int |\psi(x,y,z)|^2$ $dx\,dy\,dz < \infty$. For example, $\psi = e^{-x^2}$ is square-integrable over $-\infty \leqslant x \leqslant \infty$ but $\psi = e^{+x^2}$ is not.

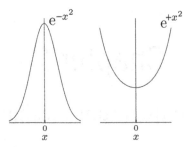

Probabilities are usually expressed on a scale from 0 (no chance) to 1 (certainty). The certainty that the particle is somewhere leads to the *normalization condition* for the wave function:

$$\int |\psi(x,y,z)|^2 \, dx\,dy\,dz = 1. \qquad (2.1)$$

Clearly, the wave function must be continuous for it to yield a physically sensible probability density. All this is summarized in:

Postulate 1. *The state of a system not subject to external time-varying forces is specified completely by a continuous and square-integrable wave function ψ that depends on the coordinates of the particles. The quantity $|\psi|^2 \, d\tau$ is the probability of finding the particles in a volume element $d\tau$ at a given location.*

2.2 Operators represent observables

In quantum chemistry, every physical observable is represented by an operator. Hence, we first study operators and then quantum chemical ones.

2.2.1 Operators

A function of one variable, like $\sin(x)$, is a 'black box' that takes any real number x as input and produces a real number as output. Similarly, an *operator* is a black box that takes a function (or vector) as input and produces a function (or vector) as output. For example, the differentiation operator d/dx takes the function $\sin(x)$ as input and produces the function $\cos(x)$ as output. A 'hat' may be used to indicate that a symbol represents an operator; for example, $\hat{A} f(x) = g(x)$ indicates that the operator $\hat{A}$ maps the function $f(x)$ on to the function $g(x)$.

We have already encountered, in equations (1.1)–(1.3), the symmetry operators C_2, σ_{xy} and i, which map a vector containing the Cartesian coordinates of a point to a vector containing the coordinates of the location to which the point is moved by the corresponding symmetry operation. A symmetry operator can act on a function by changing the arguments of the function. For example, the inversion operator changes the sign of all the arguments of a function and this can be written as $\hat{i}f(x, y, z) = f(-x, -y, -z)$.

Operators work mostly as we expect. The product of a scalar (number) c and an operator $\hat{A}$ is another operator $c\hat{A}$ defined by $[c\hat{A}]f(x) = c[\hat{A}f(x)]$. The sum of two operators is defined by $[\hat{A} + \hat{B}]f(x) = \hat{A}f(x) + \hat{B}f(x)$. The product $\hat{A}\hat{B}$ of two operators $\hat{A}$ and $\hat{B}$ is defined by $\hat{A}\hat{B}f(x) = \hat{A}[\hat{B}f(x)]$. Note that the operator on the right acts first on the function and its output is acted upon by the operator on the left. $\hat{A}^2$ is just $\hat{A}$ applied twice. In many cases, operator multiplication is *non-commutative*, that is $\hat{A}\hat{B} \neq \hat{B}\hat{A}$. For example, the operator $\hat{A} = x$ that multiplies a function by x and the differentiation operator $\hat{B} = d/dx$ do not commute:

$$\hat{A}\hat{B}f(x) = xf'(x) \neq \hat{B}\hat{A}f(x) = d/dx[xf(x)] = xf'(x) + f(x). \qquad (2.2)$$

Another example of operators that do not commute with each other is provided by σ_v and σ_v' under D_{3h} symmetry as shown in figure 1.7.

If an operator $\hat{A}$ maps a function f onto itself multiplied by a constant a, that is if

$$\hat{A}f(x) = af(x), \qquad (2.3)$$

then the function $f(x)$ is said to be an *eigenfunction* of $\hat{A}$ and the constant a is called the corresponding *eigenvalue*. For example, let $\hat{A} = d/dx$ and $f(x) = 5\,e^{3x}$. Then $\hat{A}f(x) = d(5\,e^{3x})/dx = 15\,e^{3x} = 3f(x)$ and so $5\,e^{3x}$ is an eigenfunction of the differentiation operator d/dx with eigenvalue 3.

If f is an eigenfunction of $\hat{A}$ with eigenvalue a, then so is any non-zero multiple of f. To see this, note that

$$\hat{A}(cf) = c(\hat{A}f) = c(af) = a(cf) \qquad (2.4)$$

in which c is a non-zero constant.

An operator $\hat{A}$ is *linear* if, for all functions f and g, and all constants a and b, it is true that

$$\hat{A}(af + bg) = a\hat{A}f + b\hat{A}g. \qquad (2.5)$$

Many operators, such as the differentiation operator, are linear. The square root operator is an example of an operator that is not linear. All quantum mechanical operators are linear and Hermitian. An operator $\hat{A}$ is *Hermitian* if

$$\int \phi^*(\hat{A}\psi)\,dx = \left[\int \psi^*(\hat{A}\phi)\,dx\right]^* \qquad (2.6)$$

is true for all functions ϕ and ψ. Hermitian operators have two important properties. (a) All the eigenvalues of a Hermitian operator are real numbers. (b) If $f_1(x)$ and $f_2(x)$ are any pair of distinct eigenfunctions of a Hermitian operator, then they are, or can be chosen to be, *orthogonal* to one another:

$$\int_{-\infty}^{+\infty} f_1(x)f_2(x)\,dx = 0. \qquad (2.7)$$

If the functions are complex-valued, then $f_1(x)$ should be replaced by $f_1^*(x)$ in equation (2.7).

2.2.2 Quantum chemical operators

Postulate 2. *Every observable A is represented by a linear, Hermitian operator $\hat{A}$. The operator $\hat{x}$ for each coordinate x corresponds to multiplication by x, and the operator for each component of linear momentum p_x is $\hat{p}_x = -i\hbar\partial/\partial x$ (read $\hbar$ as h-bar) in which $\hbar = h/2\pi$ and h is Planck's constant. All other operators are constructed by replacing Cartesian coordinates and linear momenta in the Newtonian formula for A by these two.*

Postulate 3. *The only values that can be observed in a measurement of an observable A are the eigenvalues of the corresponding operator $\hat{A}$.*

The eigenvalues of postulate 3 are guaranteed to be real numbers, as they must be if they are to be the results of observations, because the operators are Hermitian. The $\hat{p}_x$ operator would not be Hermitian without the constant $i = \sqrt{-1}$. Often the hat ^ is left off $\hat{x}$ and other multiplicative operators. Observe that equation (2.2) shows that $\hat{x}$ does not commute with $\hat{p}_x$.

All other quantum chemical operators are constructed from the position and momentum operators. For example, the operator $\hat{T}_x$ for the x component of the kinetic energy of a particle of mass m is obtained as follows:

$$\hat{T}_x = \frac{\hat{p}_x^2}{2m} = \frac{(-i\hbar)^2}{2m}\left(\frac{\partial}{\partial x}\right)^2 = -\frac{\hbar^2}{2m}\frac{\partial^2}{\partial x^2}. \tag{2.8}$$

The total *kinetic energy operator* for a single particle of mass m is

$$\hat{T} = \hat{T}_x + \hat{T}_y + \hat{T}_z = -\frac{\hbar^2}{2m}\nabla^2 \tag{2.9}$$

in which

$$\nabla^2 \equiv \frac{\partial^2}{\partial x^2} + \frac{\partial^2}{\partial y^2} + \frac{\partial^2}{\partial z^2}$$

is called the *Laplacian* operator or 'del-squared'. Sometimes ∇^2 is denoted by Δ (read Δ as dell-tah). The potential energy V depends only on the coordinates and so $\hat{V}$ for a single particle is simply multiplication by the potential energy function $V(x, y, z)$. The total energy operator is called the *Hamiltonian* operator:

$$\hat{H} = \hat{T} + \hat{V}. \tag{2.10}$$

The hat is usually left off $\hat{V}$ because it is a multiplicative operator. Inserting equation (2.9) into equation (2.10), we get the single-particle Hamiltonian

$$\hat{H} = -\frac{\hbar^2}{2m}\nabla^2 + V(x, y, z). \tag{2.11}$$

Hamiltonians for systems with more than one particle are discussed in section 6.2. The next section explains the fundamental role of the Hamiltonian operator.

2.3 Schrödinger's equation

Postulate 4. *The wave functions ψ of a system free of time-varying external forces are eigenfunctions of the Hamiltonian operator $\hat{H}$:*

$$\hat{H}\psi = E\psi. \tag{2.12}$$

Postulate 3 tells us that the observable energies are eigenvalues of the energy (Hamiltonian) operator. Postulate 4 tells us that the eigenfunctions of the Hamiltonian are precisely the wave functions of postulate 1.

Equation (2.12) is called the Schrödinger equation. Since it has many solutions, equation (2.12) is often written as

$$\hat{H}\psi_n = E_n\psi_n \tag{2.13}$$

where the *quantum number $n = 1, 2, \ldots$* labels the states in order of increasing energy. Since $\hat{H}$ is Hermitian, the energy eigenvalues E_n are real numbers as they must be and the eigenfunctions ψ_n are orthogonal to one another (see equation (2.7)):

$$\int \psi_m \psi_n \, \mathrm{d}\tau = 0 \qquad \text{for } m \neq n. \tag{2.14}$$

The normalization and orthogonality conditions, equation (2.1) and equation (2.14), can be combined in the compact *orthonormality condition*:

$$\int \psi_m \psi_n \, \mathrm{d}\tau = \begin{cases} 1 & \text{for } m = n, \\ 0 & \text{for } m \neq n \end{cases} \tag{2.15}$$

in which $\mathrm{d}\tau$ is the pertinent volume element; for example, $\mathrm{d}\tau = \mathrm{d}x \, \mathrm{d}y \, \mathrm{d}z$ for a single particle. If the wave functions are complex-valued, then ψ_m should be replaced by ψ_m^* in equations (2.14)–(2.15).

Equation (2.4) tells us that if ψ' is an eigenfunction of $\hat{H}$ with energy E, then so is $\psi = c\psi'$ where c is any non-zero constant. We exploit this to choose c in a manner such that ψ is normalized. We require

$$\int |\psi|^2 \, \mathrm{d}\tau = \int |c\,\psi'|^2 \, \mathrm{d}\tau = |c|^2 \int |\psi'|^2 \, \mathrm{d}\tau = 1.$$

If c is a real number, then

$$c = \pm \left(\int |\psi'|^2 \, \mathrm{d}\tau \right)^{-1/2} \tag{2.16}$$

and we can choose either the positive or the negative sign because both choices lead to the same probability density $|\psi|^2$. This is referred to as the choice of the *phase factor*. Usually the positive sign is chosen for simplicity.

For example, suppose that $\psi' = e^{-bx^2}$ is an unnormalized wave function for a single particle in one dimension and that the range of x is $(-\infty, \infty)$. Insert $\psi' = e^{-bx^2}$ into the integral in equation (2.16), note that $b > 0$ is required for square integrability, and use the integral formula (A.15) on page A-3, to find that

$$\int |\psi'|^2 \, d\tau = \int_{-\infty}^{\infty} e^{-2bx^2} \, dx = \left(\frac{\pi}{2b}\right)^{1/2}.$$

Choosing the positive root leads to the normalization constant $c = (2b/\pi)^{1/4}$ and the normalized wave function $\psi = (2b/\pi)^{1/4} e^{-bx^2}$.

2.4 Measured and average values

A way to compute the average value of an observable A that we can expect to obtain in a series of measurements on a set of identical systems is given by the next postulate.

Postulate 5. *If a system is in a state described by a normalized wave function ψ, then the average value of the observable A with corresponding operator $\hat{A}$ is given by*

$$\langle A \rangle = \int \psi(\hat{A}\psi) \, d\tau. \tag{2.17}$$

Equation (2.17) must be written as $\langle A \rangle = \int \psi^*(\hat{A}\psi) \, d\tau$ if the wave function is complex-valued. An average need not coincide with any of the numbers it is constructed from. Hence the *expectation* (or average) value $\langle A \rangle$ need not coincide with any of the eigenvalues of $\hat{A}$. The *variance* of the values of A found by measurements on a set of identical systems is given by

$$\sigma(A) = \left(\langle A^2 \rangle - \langle A \rangle^2\right)^{1/2}. \tag{2.18}$$

The Heisenberg *uncertainty principle* is:

$$\sigma(x)\sigma(p_x) \geqslant \hbar/2. \tag{2.19}$$

Heisenberg's inequality implies that $\sigma(x)$ and $\sigma(p_x)$ cannot both be zero although one can be zero if the other is infinite. In other words, we cannot simultaneously measure the position and momentum of a particle without introducing an *inherent* uncertainty in one or both of these quantities. Similar Heisenberg inequalities hold for the y and z components[2].

---

[2] Moreover, Howard P Robertson showed that uncertainty relationships hold true for all pairs of operators, $(\hat{A}, \hat{B})$, that do not commute with each other:

$$\sigma(A)\sigma(B) \geqslant \frac{1}{2}\langle|\hat{A}\hat{B} - \hat{B}\hat{A}|\rangle. \tag{2.20}$$

Inserting the quantum mechanical operators for $\hat{x}$ and $\hat{p_x}$ into equation (2.20) and doing some algebra (see equation (2.2)) leads to the Heisenberg uncertainty principle given by equation (2.19).

If quantum mechanics makes you uneasy, pay heed to Richard Feynman (Nobel Prize, 1965). He said, 'Do not keep saying to yourself: "But how can it be like that?" because you will go down the drain into a blind alley from which nobody has yet escaped. Nobody knows how it can be like that. But all known experiments back up quantum mechanics.'

Problems (see appendix B for hints and solutions)

2.1 Evaluate $g = \hat{A}f$ for (a) $\hat{A} = \mathrm{d}^2/\mathrm{d}x^2$ and $f(x) = \mathrm{e}^{-ax}$, (b) $\hat{A} = \int_0^a \mathrm{d}x$ and $f(x) = x^3 - 2x^2 + 3x - 4$, and (c) $\hat{A} = \nabla^2$ and $f(x,y,z) = x^4 y^3 z^2$.

2.2 Write down the form of the operator $\hat{A}^2$ when (a) $\hat{A} = x$, (b) $\hat{A} = \mathrm{d}/\mathrm{d}x$, and (c) $\hat{A} = \mathrm{d}/\mathrm{d}x + x$. Make sure you include a function f before carrying out operations to check your answer.

2.3 Which of the following functions (a) e^{-ax^2}, (b) $\cos \beta x$, and (c) $7\mathrm{e}^{ikx}$ are eigenfunctions of the operator $-\hbar^2 \mathrm{d}^2/\mathrm{d}x^2$? For each eigenfunction, what is the corresponding eigenvalue?

2.4 Are the functions $f(x) = \mathrm{e}^{-3x^2}$ and $g(x) = \mathrm{e}^{+3x^2}$ square integrable over $-\infty < x < \infty$? Would the answer change if the constant 3 in these functions was changed to 5 or 99?

2.5 Jan and Olga were working on solutions to the Schrödinger equation for a model one-dimensional problem. They squabbled because Jan found the ground state wave function to be $\psi = \left(\zeta^3/2\pi\right)^{1/2} \mathrm{e}^{-a|x|}$ whereas Olga found it to be $\psi = -\left(\zeta^3/2\pi\right)^{1/2} \mathrm{e}^{-a|x|}$ where a and ζ are constants. Their director, Susan, explained to them why they were both right. What was Susan's explanation?

2.6 Write down the three mathematical statements implied by 'the real-valued functions $f(x)$ and $g(x)$ are orthonormal on the interval $[0, \infty)$'.

2.7 Suppose $\psi' = \mathrm{e}^{-ar}$ is an unnormalized wave function for a single-particle, one-dimensional system. The range of r is $(0, \infty)$, and $a > 0$ as is required for square integrability. Find c such that $\psi = c\psi'$ is normalized.

Quantum Chemistry

A concise introduction for students of physics, chemistry, biochemistry and materials science

Ajit J Thakkar

Chapter 3

Translation and vibration

3.1 A particle in a wire

The postulates of quantum mechanics can be understood by considering a simple one-dimensional model of the translational motion of a particle in a wire. Many of the features encountered here, including quantization of energy levels, orthogonality of wave functions, increase in energy with number of nodes in the wave function, and symmetry of wave functions, recur throughout this book.

Consider a particle of mass m that can move freely along a straight piece of wire of length a but is prevented from leaving the wire by infinitely high walls as in figure 3.1. This model of *translational motion* is used in statistical thermodynamics when the macroscopic properties of an ideal gas are related to the properties of the molecules comprising the gas.

The potential energy is zero inside the wire and at its ends, but infinite elsewhere so that the particle cannot leave the wire. Let the wire be placed along the x axis and extend from $x = 0$ to $x = a$. Since the particle will be restricted to the x axis, this is a one-dimensional problem. $\psi(x) = 0$ for $x < 0$ and $x > a$ because the probability of

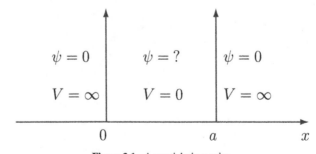

Figure 3.1. A particle in a wire.

the particle being outside the wire is zero. To find ψ in the wire, we must solve the Schrödinger equation (2.12) using a one-particle Hamiltonian, equation (2.11), with $V = 0$ and the kinetic energy operator containing only an x component. Thus, we must solve:

$$-\frac{\hbar^2}{2m}\frac{d^2\psi}{dx^2} = E\psi \qquad \text{for } 0 \leqslant x \leqslant a. \tag{3.1}$$

3.1.1 Solving the Schrödinger equation

Equation (3.1) is solved by noticing that the only functions whose second derivatives are proportional to themselves, with a negative proportionality constant, are the sine and cosine functions. So we try the function:

$$\psi(x) = A \sin \alpha x + B \cos \beta x \qquad \text{for } 0 \leqslant x \leqslant a. \tag{3.2}$$

Postulate 1 on page 2-2 says that ψ must be continuous everywhere. Continuity at the left edge ($x = 0$) requires $\psi(0) = 0$ which forces $A \times \sin 0 + B \times \cos 0 = B = 0$. Hence equation (3.2) reduces to $\psi(x) = A \sin \alpha x$ in the wire. Next, continuity at $x = a$ requires $\psi(a) = A \sin \alpha a = 0$ and hence $\alpha a = \pm n\pi$ or $A = 0$ where n is an integer. The wave function cannot be zero everywhere inside the wire and so $A = 0$ and $n = 0$ are not physically admissible. This means that $\psi_n(x) = A \sin(n\pi x/a)$ with $n \neq 0$. Changing the sign of n merely changes the sign and phase factor of ψ. Hence, the negative values of n do not lead to solutions that are physically distinct from those with positive n (see section 2.3), and it is sufficient to consider only positive n. The constant A can be found from the normalization condition (2.1). Using the integral in equation (A.9) from appendix A, we get:

$$\int_0^a |\psi_n(x)|^2 \, dx = |A|^2 \int_0^a \sin^2(n\pi x/a) \, dx = \frac{|A|^2 a}{2} = 1. \tag{3.3}$$

Hence $|A| = \sqrt{2/a}$ and we can choose $A = \sqrt{2/a}$. Finally, substitute $\psi_n(x) = A \sin(n\pi x/a)$ into the Schrödinger equation (3.1), and find

$$-\frac{\hbar^2}{2m}\frac{d^2\psi_n}{dx^2} = \frac{\hbar^2 n^2 \pi^2}{2ma^2}\psi_n = \frac{h^2 n^2}{8ma^2}\psi_n = E_n\psi_n. \tag{3.4}$$

In summary, we have found the wave functions

$$\psi_n(x) = \begin{cases} (2/a)^{1/2} \sin(n\pi x/a) & \text{inside the wire: } 0 \leqslant x \leqslant a \\ 0 & \text{outside the wire: } x < 0, x > a \end{cases} \tag{3.5}$$

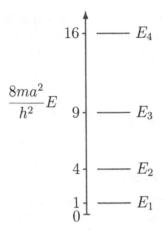

Figure 3.2. Energy levels of a particle in a wire, $E_n = h^2 n^2 / (8ma^2)$.

for $n = 1, 2, \ldots$, and from equation (3.4), the corresponding energies

$$E_n = \frac{h^2 n^2}{8ma^2} \qquad \text{for } n = 1, 2, \ldots. \tag{3.6}$$

The *quantum number* $n = 1, 2, \ldots$ labels the wave functions and energies.

3.1.2 The energies are quantized

Equation (3.6) and figure 3.2 show that only certain energies are allowed; we say that the energies are *quantized*. The lowest energy state ($n = 1$ in this case) is called the *ground state*, and the higher energy states ($n > 1$ in this case) are called *excited states*.

The lowest allowable energy is called the *zero-point energy* and is greater than zero in this problem. It implies that a quantum particle in a wire is always moving around! In this model, the Heisenberg uncertainty principle makes $E_1 = 0$ impossible because it would imply $\langle p_x^2 \rangle = \langle p_x \rangle = \sigma(p_x) = 0$ since all the energy is kinetic. That in turn would require $\sigma(x)$ to be infinite to satisfy the Heisenberg principle. However, $\sigma(x)$ cannot be infinite because the particle is confined to a wire of finite length.

The spacing between adjacent energy levels, $E_{n+1} - E_n$, increases as n increases. As the wire gets longer (a increases) or the particle gets heavier (m increases), the spacings $E_{n+1} - E_n$ get smaller and the energy levels are squished closer together. There is almost a classical energy continuum for heavy enough particles and long enough wires. The largest quantum effects are seen for extremely light particles in very short wires.

Einstein's relationship $E = h\nu$ gives $\nu = (E_u - E_l)/h$ as the frequency (read ν as new) of a photon that can excite the particle from state l to u and of a photon emitted when a particle relaxes from state u to l; see problem 3.2.

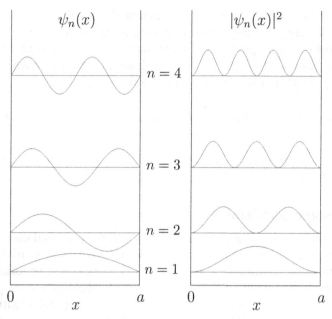

Figure 3.3. Wave functions $\psi_n(x)$ and probability densities $|\psi_n(x)|^2$ for the four lowest states of a particle in a wire.

3.1.3 Understanding and using the wave functions

The wave functions of equation (3.5) and their squares are shown in figure 3.3. The probability of finding the particle is not the same at all locations in the wire; see problem 3.4. The probability density oscillates in the excited states. In the limit as $n \to \infty$, the peaks of the oscillations are so close together that the probability density is essentially uniform inside the wire. The points inside the wire at which the wave functions cross the x axis and become zero are called *nodes* or *zero crossings*. *The energy of a state increases with the number of nodes in the corresponding wave function.*

There is a center of inversion at the midpoint of the wire, $x = a/2$. The inversion operator $\hat{i}$ interchanges the points $a/2 + \epsilon$ and $a/2 - \epsilon$ (read ϵ as ep-si-lawn) which is equivalent to the interchange of x and $a - x$. The wave functions for the states with odd n are symmetric under inversion—that is $\hat{i}\,\psi_n(x) = \psi_n(x)$ for $n = 1, 3, 5, \ldots$. The wave functions for the even n states are antisymmetric under inversion, that is $\hat{i}\,\psi_n(x) = -\psi_n(x)$ for $n = 2, 4, 6, \ldots$. The probability densities $|\psi_n(x)|^2$ are symmetric under inversion for all n as they must be since the two halves of the wire are physically indistinguishable. Hence, the probability of finding the particle in either half of the wire is $\int_0^{a/2} |\psi_n|^2 \, dx = \int_{a/2}^a |\psi_n|^2 \, dx = 1/2$. The wave functions $\psi_n(x)$ are eigenfunctions of a Hermitian operator and hence orthogonal to one another (see equation (2.14)):

$$\int_0^a \psi_n(x)\psi_m(x) \, dx = 0 \qquad \text{for } n \neq m. \tag{3.7}$$

Average values of observables can be calculated from the wave functions using postulate 5. For example, using the ground-state wave function $\psi_1(x)$ from equation (3.5), the $\hat{x}$ operator from postulate 2, equation (2.17), and the formula of equation (A.10) from appendix A to do the final integral, we find that the average value of the position of a particle in a wire in its ground state is:

$$\langle x \rangle = \int \psi_1(\hat{x}\psi_1) \, d\tau = \frac{2}{a} \int_0^a \sin(\pi x/a)x \sin(\pi x/a) \, dx = a/2.$$

The result makes sense because the probability density is symmetric with respect to the center of the wire.

3.2 A harmonic oscillator

A harmonic oscillator is a one-dimensional model for vibrational motion. Visualize it as a ball attached to a rigid surface by an ideal spring. A particle of mass m undergoing harmonic motion in the x direction experiences a restoring force proportional to its displacement. We choose $x = 0$ to be the equilibrium point of zero displacement; positive values of x correspond to stretching and negative values to compression of the spring. Then $F = -kx$ in which the force constant $k > 0$ is a measure of the spring's stiffness. Since $F = -dV/dx$, the potential is $V = kx^2/2$, and the Schrödinger equation is

$$-\frac{\hbar^2}{2m}\frac{d^2\psi}{dx^2} + \frac{1}{2}kx^2\psi = E\psi. \tag{3.8}$$

The allowed energy levels are found, by an involved derivation, to be

$$E_v = \hbar\omega(v + 1/2) \qquad \text{for } v = 0, 1, 2, \ldots \tag{3.9}$$

in which $\omega = (k/m)^{1/2}$ is the vibrational frequency (read ω as o-may-gah), and v is the quantum number. The zero-point (ground $v = 0$ state) vibrational energy is $E_0 = \hbar\omega/2$; *a quantum oscillator never stops vibrating*. The energy level spacing is constant: $E_{v+1} - E_v = \hbar\omega$; see figure 3.4. The spacing $\hbar\omega$ and zero-point energy E_0 increase as the particle mass m decreases and the force constant k increases. Lighter particles and stiffer springs vibrate faster.

Examination of figure 3.5 reveals that $\psi_v(x)$ has v nodes. As it does for a particle in a wire, the energy of a state increases as the number of nodes in the wave function increases. The wave functions are eigenfunctions of the inversion operator $\hat{i}$ that changes the sign of x. The even v states are symmetric and the odd v states are

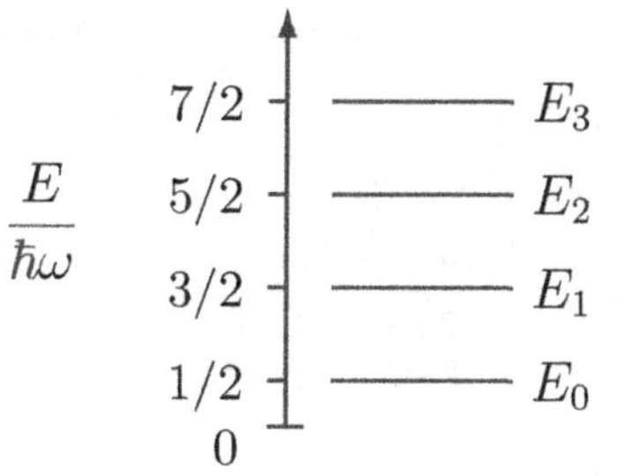

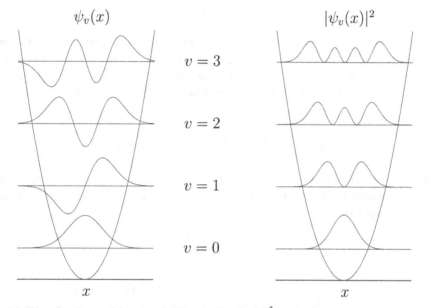

Figure 3.4. Energy levels of a harmonic oscillator.

$\psi_v(x)$ $|\psi_v(x)|^2$

$v = 3$

$v = 2$

$v = 1$

$v = 0$

x x

Figure 3.5. Wave functions $\psi_v(x)$ and probability densities $|\psi_v(x)|^2$ for the four lowest states of a harmonic oscillator.

antisymmetric with respect to $\hat{i}$; that is, $\hat{i}\,\psi_v(x) = \psi_v(x)$ for $v = 0, 2, 4, \ldots$ and $\hat{i}\,\psi_v(x) = -\psi_v(x)$ for $v = 1, 3, 5, \ldots$.

In Newtonian mechanics, an oscillator reaches its points of greatest displacement, called classical turning points, when all its energy is potential energy. Figure 3.5 shows that as v becomes large, the probability density peaks near the classical turning points. Observe in figure 3.5 that there is a non-zero probability of displacements that lie outside the classically allowed region from $x = -\sqrt{(2v+1)/\alpha}$ to $x = +\sqrt{(2v+1)/\alpha}$. This feature, called *tunneling*, is characteristic of quantum systems. Tunneling was discovered by Friedrich Hund in the wave functions for a double-well problem and first used by George Gamow to explain features of alpha decay. Quantum mechanical tunneling is important in many different phenomena, and is exploited in scanning tunneling microscopes invented by Gerd Binnig and Heinrich Rohrer (Physics Nobel Prize, 1986).

Table 3.1. Harmonic oscillator wave functions with $\alpha = m\omega/\hbar$.

$$\psi_0(x) = (\alpha/\pi)^{1/4}\, e^{-\alpha\, x^2/2}$$

$$\psi_1(x) = (4\alpha^3/\pi)^{1/4}\, x\, e^{-\alpha\, x^2/2}$$

$$\psi_2(x) = (\alpha/4\pi)^{1/4}\, (2\alpha\, x^2 - 1)\, e^{-\alpha\, x^2/2}$$

$$\psi_3(x) = (\alpha^3/9\pi)^{1/4}\, (2\alpha\, x^3 - 3x)\, e^{-\alpha\, x^2/2}$$

The wave functions for $0 \leqslant v \leqslant 3$ are listed in table 3.1. The general form of the harmonic oscillator wave functions is

$$\psi_v(x) = N_v H_v(\alpha^{1/2}x)\, e^{-\alpha\, x^2/2} \tag{3.10}$$

where $\alpha = (km)^{1/2}/\hbar = m\omega/\hbar$ and N_v is a normalization constant that can be determined using equation (2.1). The Hermite polynomial $H_v(y) = a_0 + a_1 y + a_2 y^2 + \cdots + a_v y^v$ in which $y = \sqrt{\alpha}x$ and the a_i are constants. The wave functions are eigenfunctions of a Hermitian operator and hence form an orthonormal set:

$$\int_{-\infty}^{\infty} \psi_v(x)\psi_{v'}(x)\, dx = \begin{cases} 1 & \text{for } v = v' \\ 0 & \text{for } v \neq v'. \end{cases} \tag{3.11}$$

Average values can be calculated using equation (2.17), wave functions from table 3.1, and integral formulas from appendix A. For example, the average value of x^2 in the ground state is:

$$\langle x^2 \rangle = \int_{-\infty}^{\infty} \psi_0(x)[x^2\, \psi_0(x)]\, dx$$

$$= \left(\frac{\alpha}{\pi}\right)^{1/2} \int_{-\infty}^{\infty} e^{-\alpha\, x^2/2}\, x^2\, e^{-\alpha\, x^2/2}\, dx = 1/(2\alpha).$$

The probability of non-classical displacements can be calculated to be 0.16 in the ground state. This probability decreases as v increases.

3.2.1 Molecular vibrations

Molecular vibrations are described by the harmonic oscillator model. A diatomic molecule is modeled as two atoms attached to each other by an ideal spring. The potential energy depends only on the position of one atom relative to that of the other. A center-of-mass transformation allows one to separate external from internal coordinates in the two-particle Schrödinger equation. Only the internal vibrational motion is of interest, since the external motion is simply the translational motion of the molecule as a whole. The Schrödinger equation for the vibrational motion differs from the Schrödinger equation for the harmonic oscillator only in the replacement of the mass m by the reduced mass $\mu = m_1 m_2/(m_1 + m_2)$ (read μ as mew) of the molecule where m_1 and m_2 are the masses of the two atoms. Hence, the harmonic oscillator wave functions and energy levels, with m replaced by μ, describe the stretching vibration of a diatomic molecule.

It requires $3N$ coordinates to specify the positions of the atoms in a molecule with N atoms. Three coordinates are used to specify the location of its center of mass or, in other words, to describe the translational motion of the molecule as a whole. The displacements of the remaining $3N - 3$ coordinates (called the remaining $3N - 3$ degrees of freedom) describe the internal motions of the molecule. A non-linear molecule has three independent rotations, one about each of the three axes in a coordinate system fixed at its center of mass. However, a linear molecule has only two distinct rotations, one about each of the two axes perpendicular to the molecular (z) axis. One may also think of the number of independent rotations as the number of angles needed to describe the orientation of the molecule relative to the coordinate system at its center of mass. All the remaining degrees of freedom describe molecular vibrations. Therefore, a molecule has $3N - 6$ distinct vibrations if it is non-linear or $3N - 5$ if it is linear. Each vibration can be modeled as a harmonic oscillator with different parameters.

Infrared and other types of spectroscopy enable us to measure the frequencies of transitions between vibrational energy levels of a molecule. The measured frequency can be used to deduce the force constant using the harmonic oscillator model. The frequency ν is related to the wavelength λ by $\nu = c/\lambda$, where c is the speed of light. Moreover, ν is related to the energy-level spacing by $h\nu = \Delta E$. The harmonic oscillator model predicts $\Delta E = \hbar\omega$. Combining these relationships gives us $\omega = 2\pi\nu = 2\pi c/\lambda$. Since $\omega = (k/\mu)^{1/2}$, it follows that $k = 4\pi^2 c^2 \mu/\lambda^2$. For example, the 'fundamental' $v = 0 \rightarrow 1$ vibrational transition in $^{35}Cl_2$ was observed at 565 cm^{-1}; that is, $1/\lambda = 565$ cm^{-1}. The atomic mass of ^{35}Cl is 34.969 u and so $\mu = 17.4845\,u$ for $^{35}Cl_2$. Converting all quantities to SI units and substituting into $k = 4\pi^2 c^2 \mu/\lambda^2$ gives $k = 329$ N m^{-1}. Force constants for a wide variety of molecular vibrations have been obtained in this manner. Force constants give us an indication of the relative stiffness of bonds.

Problems (see appendix B for hints and solutions)

3.1 What would happen to the energy level spacing of a particle in a wire if its mass was halved?

3.2 Calculate the wavelength of the photon emitted when an electron in a wire of length 500 pm drops from the $n = 2$ level to the $n = 1$ level.

3.3 Consider a particle in a wire. For which states (values of n) are the probabilities of finding the particle in the four quarters of the wire ($0 \leqslant x \leqslant a/4$, $a/4 \leqslant x \leqslant a/2$, $a/2 \leqslant x \leqslant 3a/4$, and $3a/4 \leqslant x \leqslant a$) all equal to $1/4$? Explain your reasoning with sketches of pertinent $|\psi_n(x)|^2$.

3.4 Consider the ground state of a particle in a wire. Calculate the probability of finding the particle in (a) the left half of the wire, and (b) in each quarter of the wire.

3.5 Calculate the ground-state expectation values $\langle x \rangle$, $\langle x^2 \rangle$, $\langle p_x \rangle$, and $\langle p_x^2 \rangle$ for a particle in a wire. Explain how the result for $\langle p_x^2 \rangle$ makes physical sense. Then calculate $\sigma(x)\sigma(p_x)$ and check whether the Heisenberg uncertainty principle is satisfied for the ground state of a particle in a wire.

3.6

(a) The particle in a wire can be used as a very simple model of the π electrons in 1,3-butadiene. Assume that there can be no more than two electrons per energy level on the basis of the Pauli principle that you learned in your general chemistry course. Choose m and a in a meaningful manner. Then write down a formula for the wavelength of a photon that has precisely the energy required to induce the promotion of an electron from the highest filled to the lowest unfilled energy level. Finally, find a numerical value for this wavelength.

(b) Solve the same problem for 1,3,5-hexatriene.

(c) Generalize the solution to conjugated polyenes with $2N$ carbon atoms.

3.7 What would happen to the energy level spacing of a harmonic oscillator if its force constant was multiplied by four?

3.8 Calculate the ground-state expectation values $\langle x \rangle$ and $\langle x^2 \rangle$ for a one-dimensional harmonic oscillator. Then calculate $\sigma(x)$.

3.9 An experimental measurement of the $v = 0 \rightarrow 1$ vibrational transition enables us to calculate the force constant. Vibrational transitions are usually given in wave numbers (reciprocal wavelengths): $\nu/c = 1/\lambda$. The wave numbers for $^1H^{35}Cl$, $^1H^{81}Br$, and $^1H^{127}I$ were found to be 2988.9, 2649.7, and 2309.5 cm^{-1}, respectively. The atomic masses for ^1H, ^{35}Cl, ^{81}Br, and ^{127}I are 1.0078, 34.969, 80.916, and 126.90 u, respectively. Use the harmonic oscillator model of molecular vibrations to calculate the force constants in N m^{-1} (equivalent to J m^{-2}) and discuss the relative stiffness of the bonds.

Quantum Chemistry
A concise introduction for students of physics, chemistry, biochemistry and materials science
Ajit J Thakkar

Chapter 4

Symmetry and degeneracy

4.1 A particle in a rectangular plate

We noted the link between molecular symmetry and polarity in section 1.3. We saw symmetry in the wave functions for the model systems discussed in chapter 3. Here, we study translational motion in two dimensions as a prelude to a systematic investigation of how symmetry affects energy levels in section 4.2, probability densities in section 4.3, and wave functions in section 4.4 and section 4.5.

Consider a particle of mass m allowed to move freely within a rectangular plate, $0 \leqslant x \leqslant a$, $0 \leqslant y \leqslant b$, but trapped in it by an infinite potential energy outside the plate as shown in figure 4.1.

The wave function $\psi = 0$ everywhere outside the plate. Within the plate, $V = 0$ and ψ is determined by the two-dimensional Schrödinger equation:

$$-\frac{\hbar^2}{2m}\frac{\partial^2 \psi_{n_x,n_y}}{\partial x^2} - \frac{\hbar^2}{2m}\frac{\partial^2 \psi_{n_x,n_y}}{\partial y^2} = E_{n_x,n_y}\psi_{n_x,n_y}. \tag{4.1}$$

The two-dimensional problem requires two quantum numbers.

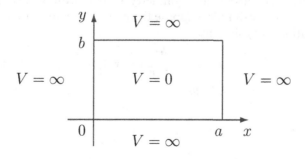

Figure 4.1. A particle in a rectangular plate.

doi:10.1088/978-1-627-05416-4ch4

A partial differential equation like equation (4.1) is often difficult to solve. However, in this case the technique of *separation of variables* can be used. Suppose that the Hamiltonian for a two-dimensional problem is the sum of two one-dimensional Hamiltonians $\hat{H} = \hat{H}_x + \hat{H}_y$ which act *only* on x and y, respectively, and that their eigenvalues and eigenfunctions are known:

$$\hat{H}_x \varphi_j(x) = \varepsilon_j \varphi_j(x) \qquad \text{for } j = 1, 2, \ldots \tag{4.2}$$

$$\hat{H}_y \phi_k(y) = \epsilon_k \phi_k(y) \qquad \text{for } k = 1, 2, \ldots. \tag{4.3}$$

Then the eigenfunctions of the two-dimensional Schrödinger equation

$$\hat{H} \psi_{j,k}(x, y) = E_{j,k} \psi_{j,k}(x, y) \tag{4.4}$$

are given by $\psi_{j,k}(x, y) = \varphi_j(x) \phi_k(y)$ and the corresponding eigenvalues are $E_{j,k} = \varepsilon_j + \epsilon_k$.

The Hamiltonian for a particle in a rectangle is separable into two one-dimensional Hamiltonians both corresponding to a particle in a wire, see section 3.1. Hence, the energies for the particle in a rectangle are

$$E_{n_x, n_y} = \frac{h^2}{8m} \left(\frac{n_x^2}{a^2} + \frac{n_y^2}{b^2} \right) \qquad \text{for } n_x = 1, 2, \ldots \text{ and } n_y = 1, 2, \ldots \tag{4.5}$$

and the corresponding wave functions are

$$\psi_{n_x, n_y}(x, y) = \begin{cases} (4/ab)^{1/2} \sin(n_x \pi x / a) \sin(n_y \pi y / b) & \text{inside the plate,} \\ 0 & \text{outside the plate.} \end{cases} \tag{4.6}$$

Examine the wave functions shown in figure 4.2 for the four lowest-energy states. Observe that the ground-state wave function $\psi_{1,1}$ does not cross zero inside the plate; that is, $\psi_{1,1}$ has no interior nodes. However, the excited state wave functions are zero at every point along certain lines, called *nodal lines*, inside the plate. Notice from figure 4.2 that $\psi_{2,1}$ has a nodal line along $x = a/2$, that $\psi_{1,2}$ has a nodal line along $y = b/2$, and that $\psi_{2,2}$ has two nodal lines, one along $x = a/2$ and the other along $y = b/2$. The energy increases with the number of nodal lines but states, like $\psi_{1,2}$ and $\psi_{2,1}$, with the same number of nodal lines may have different energies.

4.2 Symmetry leads to degeneracy

Consider the effect of introducing symmetry in the two-dimensional rectangular plate of section 4.1. For example, suppose the plate is square so that $b = a$; then the energies of equation (4.5) simplify to

$$E_{n_x, n_y} = \frac{h^2}{8ma^2} \left(n_x^2 + n_y^2 \right). \tag{4.7}$$

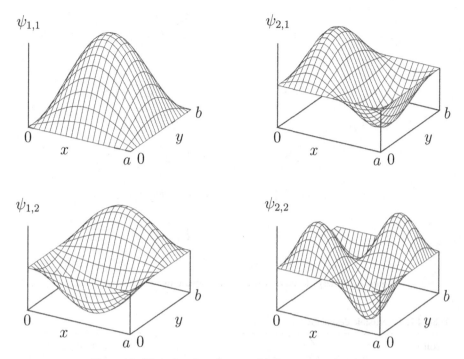

Figure 4.2. Wave functions for a particle in a rectangular plate.

The two lowest excited state energies are equal:

$$E_{2,1} = E_{1,2} = 5h^2/(8ma^2) \tag{4.8}$$

although the corresponding wave functions are different. Each different wave function corresponds to a different state but more than one state may have the same energy (or energy level). An energy level that corresponds to more than one state is called *degenerate*, with a *degeneracy* equal to the number of corresponding states. Thus the second lowest energy level of a particle in a square two-dimensional plate, $E = 5h^2/(8ma^2)$, is two-fold degenerate, whereas the ground state has a degeneracy of one and is said to be non-degenerate. Many two-fold degeneracies arise for the particle in a square plate because $E_{n,k} = E_{k,n}$ for all $k \neq n$. Indeed, a particle in a square plate has more degenerate energy levels than non-degenerate ones, as seen clearly in figure 4.3.

Degeneracies almost always arise from some symmetry of the system. For example, a particle in a square plate has C_{4v} symmetry, and one finds symmetry-induced degeneracies in the energy levels. The allowed degeneracies for all the point groups are listed in table 4.1. There are no degeneracies induced by the geometrical symmetry if the point group is Abelian (see section 1.3). A C_n axis with $n \geqslant 3$ is usually a good indicator of degeneracy with exceptions only for the rarely encountered C_n and C_{nh} groups.

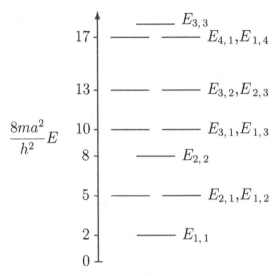

Figure 4.3. Energy levels of a particle in a square two-dimensional plate.

Table 4.1. Permissible degeneracies.

Group	Maximum degeneracy
C_1, C_i, C_s, C_{2v}, D_2, D_{2h}, and C_n, C_{nh}, S_{2n} for $n = 2, 3, \ldots$	1
D_{2d}, and C_{nv}, D_n, D_{nd}, D_{nh} for $n = 3, 4, \ldots$	2
T, T_h, T_d, O, O_h	3
I, I_h	5
K (sphere)	∞

Sometimes there are *accidental degeneracies*. For example, consider the energies of a particle in a square, equation (4.7). When (n_x, n_y) differs from $(\bar{n}_x, \bar{n}_y)$ by more than an interchange, we expect $n_x^2 + n_y^2 \neq \bar{n}_x^2 + \bar{n}_y^2$ and hence $E_{n_x, n_y} \neq E_{\bar{n}_x, \bar{n}_y}$. However, there are some rare cases where equality holds. For example, $5^2 + 5^2 = 1^2 + 7^2$ leads to the accidental degeneracy $E_{5,5} = E_{1,7}$ and $1^2 + 8^2 = 4^2 + 7^2$ leads to $E_{1,8} = E_{4,7}$.

There can also be degeneracies induced by a *non-geometrical symmetry*; for example, such degeneracies arise in the hydrogen atom, as will be seen in section 6.3. Understanding non-geometrical symmetry is beyond the scope of this book.

4.3 Probabilities in degenerate states

How do degenerate states differ from one another? Consider the wave functions for a particle in a square plate. Setting $b = a$ in equation (4.6) gives:

$$\psi_{n_x, n_y}(x, y) = \left(\frac{2}{a}\right) \sin\left(\frac{n_x \pi x}{a}\right) \sin\left(\frac{n_y \pi y}{a}\right). \tag{4.9}$$

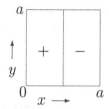 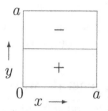

Figure 4.4. Sign patterns and nodal lines of $\psi_{2,1}$ (left) and $\psi_{1,2}$ (right).

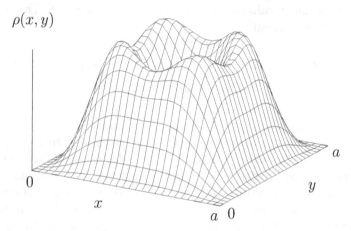

Figure 4.5. The ensemble-averaged probability density for the two-fold degenerate $\psi_{1,2}$ and $\psi_{2,1}$ states.

The wave function for the 2,1 state equals zero along the line $x = a/2$, whereas the wave function for the 1,2 state equals zero along the line $y = a/2$ as indicated in figure 4.4.

The two states are related by a rotation of 90° about the C_4 axis of symmetry passing through the center of the square, $(a/2, a/2)$, and perpendicular to it; see figure 4.2 and figure 4.4. The 90° rotation should leave the system in a configuration that is indistinguishable from the first. Hence, the two states should have the same energy and they do. However, neither $|\psi_{1,2}|^2$ nor $|\psi_{2,1}|^2$ can be a valid probability density because the two are related by a C_4 rotation whereas the probability density is an observable quantity and should not change when a symmetry operation is carried out. The solution is to notice that when there is a degeneracy, the system is equally likely to be found in any one of the degenerate states, and hence *the observable probability density is the ensemble-average of the probability densities of the degenerate states*. Thus, the observable probability density for the two-fold degenerate $\psi_{1,2}$ and $\psi_{2,1}$ states is (read ρ as row):

$$\rho(x, y) = \frac{1}{2}\left(|\psi_{1,2}|^2 + |\psi_{2,1}|^2\right). \tag{4.10}$$

Figure 4.5 shows that $\rho(x, y)$ is invariant to a C_4 rotation and has C_{4v} symmetry as it should. There is a node at the center of the square, four equivalent maxima at

(0.3, 0.3), (0.3, 0.7), (0.7, 0.7), and (0.7, 0.3) that are related by C_4 rotations, and four equivalent saddle points at $(a/2, 0.3)$, $(a/2, 0.7)$, $(0.7, a/2)$, and $(0.3, a/2)$ that are related by C_4 rotations.

4.4 Are degenerate wave functions unique?

Suppose that ψ and φ (read φ as fye) are a pair of degenerate eigenfunctions of the Hamiltonian $\hat{H}$ so that $\hat{H}\psi = E\psi$ and $\hat{H}\varphi = E\varphi$. Then any linear combination (mixture) of ψ and φ, such as $f = a\psi + b\varphi$ where the constants a and b are mixing coefficients, is also an eigenfunction of $\hat{H}$ with the same energy E. This follows from the linearity of $\hat{H}$ (see equation (2.5)):

$$\hat{H}f = \hat{H}(a\psi + b\varphi) = a\hat{H}\psi + b\hat{H}\varphi = aE\psi + bE\varphi$$
$$= E(a\psi + b\varphi) = Ef. \tag{4.11}$$

In other words, *degenerate wave functions are not unique*.

A pair of doubly degenerate wave functions (ψ, φ) can be replaced by a different pair of functions, f as defined above and $g = c\psi + d\varphi$, *provided* that we take care to choose the mixing coefficients, a, b, c, and d, such that f is normalized, g is normalized, and that f and g are orthogonal to each other. One degree of freedom is left after these three requirements on four constants are satisfied. The requirements are met by choosing $a = d = \cos\Omega$ and $b = -c = \sin\Omega$—that is, by requiring the transformation to be *unitary*. Note that $\Omega = 0°$ leads to no change at all ($f = \psi$, $g = \varphi$), and that $\Omega = 90°$ leads to ($f = \varphi$, $g = -\psi$), which amounts to no more than changing the sign of one of the functions. This transformation of degenerate wave functions is akin to rotating the axes of a coordinate system by an angle Ω while preserving the orthogonality and normalization of the basis vectors. Both pairs of coordinate axes in figure 4.6 are equally valid but one set may be more useful than the other in a particular problem. In some problems, transformations among degenerate wave functions can be utilized to make some aspects easier to work with. Examples of this will be mentioned in chapter 5.

Because the transformation between equivalent sets of degenerate wave functions is unitary and preserves orthonormality, *all equivalent sets of degenerate wave*

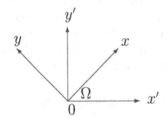

Figure 4.6. Rotation of coordinate axes.

functions lead to the same ensemble-averaged probability density. For example, for the f and g defined above, it is easy to prove that

$$\rho = \frac{1}{2}\left(|f|^2 + |g|^2\right) = \frac{1}{2}(|\psi|^2 + |\varphi|^2).\qquad(4.12)$$

4.5 Symmetry of wave functions

What do symmetry operations do to wave functions? A symmetry operation carries a molecule into a configuration that is physically indistinguishable from the original configuration. Therefore, the energy of a molecule must be the same before and after a symmetry operation $\hat{O}$ is carried out on it. Hence, the Hamiltonian (energy) operator $\hat{H}$ commutes with each of the symmetry operators $\hat{O}$; that is, $\hat{H}\hat{O} = \hat{O}\hat{H}$. It follows that

$$\hat{H}(\hat{O}\psi) = \hat{O}\hat{H}\psi = \hat{O}E\psi = E(\hat{O}\psi).\qquad(4.13)$$

Comparison of the first and last terms above shows that $\hat{O}\psi$ *is an eigenfunction of the Hamiltonian with the eigenvalue E.* Thus, we see that both ψ and $\hat{O}\psi$ are eigenfunctions of $\hat{H}$ with the same eigenvalue E.

If the energy level E is non-degenerate, then ψ is the only eigenfunction of the Hamiltonian with that eigenvalue, and it follows that $\hat{O}\psi$ can differ from ψ by at most a phase factor, see section 2.3. This usually means that $\hat{O}\psi = \pm\psi$. In other words, the wave function is an eigenfunction, typically with an eigenvalue of $+1$ or -1, of each of the symmetry operators $\hat{O}$. In any case, the effect of $\hat{O}$ on ψ is to multiply it by a number of unit magnitude, and hence $\hat{O}$ leaves the probability density $|\psi|^2$ unchanged as it should. We have already noted in section 3.1 that the wave functions for a particle in a wire with even and odd n are eigenfunctions of $\hat{i}$ with eigenvalues -1 and $+1$, respectively. Analogously, in section 3.2, we observed that the wave functions for a harmonic oscillator with even and odd v are eigenfunctions of $\hat{i}$ with eigenvalues $+1$ and -1, respectively. The apparent difference between the two systems arises because the quantum numbers begin at $n = 1$ for the particle in a wire but at $v = 0$ for a harmonic oscillator.

What about the symmetry of a set of g degenerate wave functions? The conclusion from the symmetry argument of equation (4.13)—$\hat{O}\psi$ is an eigenfunction of the Hamiltonian with eigenvalue E—tells us that $\hat{O}\psi$ could be any one of the g degenerate wave functions that correspond to the energy level E or any linear combination (mixture) thereof. In other words, any symmetry operation of the symmetry group of the system takes one member of a set of degenerate wave functions into another member of that set *or* a mixture of the members of that degenerate set. Therefore, we cannot talk about the symmetry of a single member of a degenerate set of wave functions. Instead, we must *always consider the symmetry of an entire set of degenerate wave functions.*

Problems (see appendix B for hints and solutions)

4.1 Using the particle in a square ($a = b$), two-dimensional, plate as a model, but without doing calculations, sketch the energy-level diagram for the π electrons in cyclobutadiene. Assume that there can be no more than two electrons per energy level as you expect on the basis of the Pauli principle that you learned in your general chemistry course. Use the energy-level diagram to explain the observation that it has not proved possible to isolate cyclobutadiene (C_4H_4).

4.2 Use sketches to show the regions of different sign in the degenerate wave functions $\psi_{2,3}$ and $\psi_{3,2}$ for a particle in a square two-dimensional plate.

4.3 Use the separation of variables technique (section 4.1) to write down the energy levels of a two-dimensional harmonic oscillator with mass m and force constants k_x and k_y in the x and y directions, respectively. The Hamiltonian is

$$-\frac{\hbar^2}{2m}\frac{\partial^2}{\partial x^2} - \frac{\hbar^2}{2m}\frac{\partial^2}{\partial y^2} + \frac{1}{2}k_x x^2 + \frac{1}{2}k_y y^2.$$

What happens if $k_y = k_x$?

4.4 The separation of variables technique also works for two groups of variables; for example, (x, y, z) could be separated into (x, y) and z. Apply this idea and use the results of section 3.1 and section 4.1 to write down the wave functions and energy levels of a particle of mass m confined to a three-dimensional box, $0 \leqslant x \leqslant a$, $0 \leqslant y \leqslant b$, $0 \leqslant z \leqslant c$, by an infinite potential energy outside the box but allowed to move freely within it.

4.5 What happens if the box in problem 4.4 is a cube of side length a? Write down the energy level formula. What are the quantum numbers of all the states that correspond to the first excited energy level?

4.6 Decide which of the molecules in problem 1.6 will have degeneracies in their energy levels.

4.7 Decide which of the molecules in problem 1.7 will have degeneracies in their energy levels.

4.8 Explain why the ground state of an 18-electron molecule may have unpaired electrons if it belongs to the D_{6h} point group, whereas it must have all its electrons paired if it belongs to the D_{2h} point group.

4.9 The inversion operator $\hat{i}$ for a particle in a wire stretching from $x = 0$ to $x = a$ interchanges the points x and $a - x$; in other words, $\hat{i}\psi_n(x) = \psi_n(a - x)$. Prove that

$$\hat{i}\psi_n(x) = \begin{cases} +\psi_n(x) & \text{for } n = 1, 3, 5, \ldots \\ -\psi_n(x) & \text{for } n = 2, 4, 6, \ldots \end{cases}$$

Quantum Chemistry
A concise introduction for students of physics, chemistry, biochemistry and materials science
Ajit J Thakkar

Chapter 5

Rotational motion

5.1 A particle on a ring

In this chapter, we examine the prototypical models for rotational motion. These models provide the foundation for rotational spectroscopy, and the wave functions also play an important role in the electronic structure of atoms and diatomic molecules.

Consider a particle of mass M constrained, by infinite potentials, to move on the perimeter of a circle of radius R centered at the origin in the xy plane.

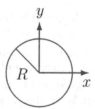

On the ring, the potential energy is zero and the Schrödinger equation is

$$-\frac{\hbar^2}{2M}\left(\frac{\partial^2\psi}{\partial x^2}+\frac{\partial^2\psi}{\partial y^2}\right) = E\psi. \tag{5.1}$$

This two-dimensional equation can be reduced to one dimension by transforming from Cartesian to plane polar coordinates. Figure 5.1 shows that the plane polar coordinates of a point P consist of its distance r from the origin O, and the angle $\phi = \angle xOP$ (read ϕ as fye) between the x axis and the line OP. The particle is confined to the perimeter of the circle and hence its radial coordinate remains fixed at $r = R$. Thus, the angle ϕ suffices to determine the position of the particle and the wave function depends only upon ϕ.

doi:10.1088/978-1-627-05416-4ch5

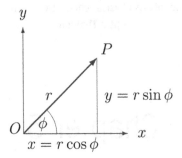

Figure 5.1. Plane polar coordinates. The point P has the Cartesian coordinates (x, y) and the plane polar coordinates (r, ϕ).

Repeated application of the chain rule allows us to express the partial derivatives with respect to x and y in equation (5.1) in terms of partial derivatives with respect to r and ϕ. Since r is fixed at R, the derivatives with respect to r vanish, and one is left with the one-dimensional equation:

$$-\frac{\hbar^2}{2I}\frac{d^2\psi}{d\phi^2} = E\psi \tag{5.2}$$

where $I = MR^2$ is the moment of inertia.

Solving equation (5.2) leads to the energy levels:

$$E_m = \frac{\hbar^2 m^2}{2I} \qquad \text{with } m = 0, \pm 1, \pm 2, \ldots . \tag{5.3}$$

The ground state ($m = 0$) is non-degenerate but the excited states are doubly degenerate because the energies of equation (5.3) are the same for the m and $-m$ states. The degeneracies are due to a $C_\infty(z)$ symmetry axis; see section 4.2. The physical meaning of the degeneracies is easy to understand. Classically, the energy of a rotating particle is related to its *angular momentum* L by $E = L^2/(2I)$. Moreover, the angular momentum of the particle on the ring is all directed in the z direction, perpendicular to the ring. Hence, we can deduce from equation (5.3) that $L_z^2 = \hbar^2 m^2$ and that $L_z = \hbar m$. Note that $\hbar$ has the dimensions of angular momentum. We see that the angular momenta of a pair of states with the same value of $|m| \geqslant 1$ are equal in magnitude but opposite in sign. Therefore the difference between two states with the same $|m| \geqslant 1$ is the direction of motion, clockwise or counterclockwise, of the particle around the ring.

$$m < 0 \quad m > 0$$

Such pairs of states clearly should have the same energy and they do.

A particle on a ring has no zero-point energy: $E_0 = 0$. The particle stops moving in the ground state unlike a particle in a wire and a harmonic oscillator. Most, but not all, systems have no rotational zero-point energy. Note from figure 5.2 that

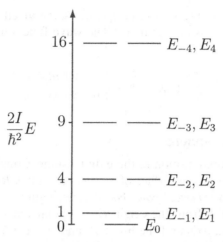

Figure 5.2. Energy levels of a particle on a ring.

the spacing between the levels, $\Delta E_m = E_{|m|+1} - E_{|m|}$, increases as $|m|$ increases. Equation (5.3) shows that $\Delta E_m = \hbar^2(2|m| + 1)/(2I)$; hence, the level spacing increases as the moment of inertia decreases, which happens when either the particle mass or the radius of the ring decreases. In other words, *light particles on small rings are the most quantum*.

The wave functions are found to be:

$$\psi_m(\phi) = \begin{cases} \dfrac{1}{\sqrt{2\pi}} & \text{for } m = 0 \\[2ex] \dfrac{1}{\sqrt{\pi}}\cos(m\phi) & \text{for } m = 1, 2, \ldots \\[2ex] \dfrac{1}{\sqrt{\pi}}\sin(m\phi) & \text{for } m = -1, -2, \ldots. \end{cases} \tag{5.4}$$

Recall from section 4.4 that the above wave functions for $m > 0$ are not unique[1] since the states with $\pm m$ are degenerate. The probability density must be computed as an average over the degenerate states (see section 4.3). Using equation (5.4) and the identity $\sin^2 y + \cos^2 y = 1$, the ensemble-averaged probability density for the $\psi_{\pm m}$ states is found to be:

$$\rho = \frac{1}{2}\left(|\psi_m|^2 + |\psi_{-m}|^2\right) = \frac{1}{2\pi}\left[\cos^2(m\phi) + \sin^2(m\phi)\right] = \frac{1}{2\pi}. \tag{5.5}$$

---

[1] In some problems, it is more convenient to work with complex-valued forms of these wave functions but we will not deal with them in this book.

So the probability of finding the particle is the same at all points on the ring as expected from the C_∞ axis of symmetry. The wave functions are orthonormal, as expected from equation (2.15):

$$\int_0^{2\pi} \psi_m \psi_n \, d\phi = \begin{cases} 1 & \text{for } m = n \\ 0 & \text{for } m \neq n. \end{cases} \tag{5.6}$$

5.2 A particle on a sphere

Next we turn to rotational motion in three dimensions. Consider a particle of mass M constrained to move on the surface of a sphere of radius R centered at the origin.

It is best to work in *spherical polar coordinates*, figure 5.3, which consist of the distance r of the point P from the origin O, the polar angle $\theta = \angle zOP$ (read θ as thay-tah) between the z axis and the line OP, and the azimuthal angle $\phi = \angle xOQ$ between the x axis and OQ, the projection of OP upon the xy plane. The two angles give the location of a point on the surface of a sphere of radius r. The polar angle is a colatitude; the point P moves from the North pole to the South pole as θ sweeps over its full range from $\theta = 0$ to $\theta = \pi$. The azimuthal angle plays the role of a longitude and has the range $0 \leqslant \phi \leqslant 2\pi$. The radius r cannot be negative and has the range $0 \leqslant r < \infty$. Note that

$$r^2 = x^2 + y^2 + z^2. \tag{5.7}$$

The volume elements in Cartesian and spherical polar coordinates are related by the *Jacobian* of the transformation, $r^2 \sin \theta$, as follows;

$$dx \, dy \, dz = r^2 \sin \theta \, dr \, d\theta \, d\phi. \tag{5.8}$$

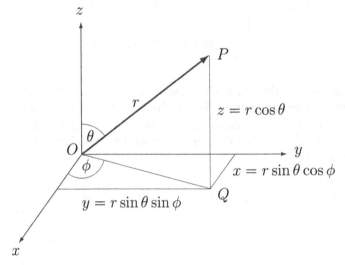

Figure 5.3. Spherical polar coordinates. The point P has Cartesian coordinates (x, y, z) and spherical polar coordinates (r, θ, ϕ).

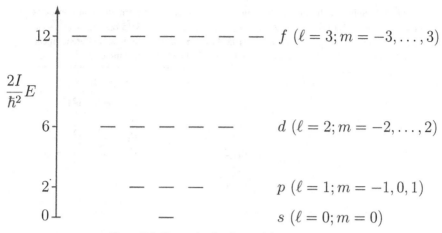

Figure 5.4. Energy levels of a particle on a sphere.

Since the particle is on the surface of a sphere, $r = R$ is a constant and the wave function depends only on the angles (θ, ϕ). The Schrödinger equation is solved by separation of variables. The ϕ equation turns out to be the same as that for a particle on a ring, but the θ equation is complicated.

This two-dimensional problem has two quantum numbers ℓ and m restricted to the values $\ell = 0, 1, 2, \ldots$ and $m = 0, \pm 1, \pm 2, \ldots, \pm \ell$ or, more concisely, $l \geqslant |m|$. The energies depend only on ℓ as follows:

$$E_{\ell,m} = \frac{\hbar^2}{2I} \ell(\ell + 1) \qquad \text{for } \ell = 0, 1, \ldots \tag{5.9}$$

where $I = MR^2$ is the moment of inertia. As seen in figure 5.4, degenerate energy levels occur (see section 4.2) because a sphere has the highest possible symmetry. The degeneracy of the energy level with quantum number ℓ is $2\ell + 1$. For example, the $\ell = 2$ energy level is five-fold degenerate and the $\ell = 2$ states with $-2 \leqslant m \leqslant 2$ all have the same energy. There is no zero-point rotational energy ($E_{0,0} = 0$) as in most, but not all, systems involving rotational motion. The level spacing increases with ℓ. The spacing increases when either the particle mass or the radius of the sphere decreases. Light particles on tiny spheres are the most quantum. Classically, the energy of a rotating particle is related to its *angular momentum* L by $E = L^2/(2I)$. Comparison with equation (5.9) reveals that the square of the angular momentum is $\langle \hat{L}^2 \rangle = \ell(\ell + 1)\hbar^2$. The azimuthal quantum number m determines the z component of the angular momentum: $\langle L_z \rangle = m\hbar$. In this way, the orientation of a rotating body is quantized—that is, it is restricted to particular orientations. This is called 'space quantization'. The other two components of the angular momentum, L_x and L_y, do not have sharply determined values simultaneously because of the uncertainty principle[2].

---

[2] Each of the operators for the three Cartesian components $\hat{L}_x$, $\hat{L}_y$, and $\hat{L}_z$ commutes with the $\hat{L}^2$ operator but $\hat{L}_x$, $\hat{L}_y$, and $\hat{L}_z$ do not commute with each other. Hence, if one component is specified exactly then the others are completely unspecified because of Robertson's uncertainty principle; see equation (2.20) in the footnote on page 2-6.

Table 5.1. Spherical harmonics $\mathcal{Y}_{\ell,m}$ in Cartesian form. The normalization constants are $c_0 = r^{-\ell}\sqrt{1/(4\pi)}$, $c_1 = \sqrt{3}\,c_0$, $c_2 = \sqrt{15/4}\,c_0$, $c_3 = \sqrt{15}\,c_0$, $c_4 = \sqrt{5/4}\,c_0$, $d_0 = \sqrt{7/4}\,c_0$, $d_1 = \sqrt{21/8}\,c_0$, $d_2 = \sqrt{105/4}\,c_0$, and $d_3 = \sqrt{35/8}\,c_0$. Different linear combinations of the f harmonics are used when dealing with O_h symmetry.

m	s ($\ell = 0$)	p ($\ell = 1$)	d ($\ell = 2$)	f ($\ell = 3$)
3				$d_3\, x(x^2 - 3y^2)$
2			$c_2\,(x^2 - y^2)$	$d_2\, z(x^2 - y^2)$
1		$c_1\, x$	$c_3\, xz$	$d_1\, x(4z^2 - x^2 - y^2)$
0	c_0	$c_1\, z$	$c_4\,(2z^2 - x^2 - y^2)$	$d_0\, z(2z^2 - 3x^2 - 3y^2)$
-1		$c_1\, y$	$c_3\, yz$	$d_1\, y(4z^2 - x^2 - y^2)$
-2			$c_3\, xy$	$d_2\, xyz$
-3				$d_3\, y(3x^2 - y^2)$

5.2.1 Rotational wave functions

The wave functions for this problem, called *spherical harmonics* by mathematicians, can be written as $\Theta_{\ell,|m|}(\cos\theta)\,\Phi_m(\phi)$ where the $\Theta_{\ell,|m|}(\cos\theta)$ are polynomials[3] in $\cos\theta$, and the $\Phi_m(\phi)$ are the wave functions given in equation (5.4) for the particle on a ring. Trigonometric identities and the relationships in figure 5.3 are used to obtain the Cartesian forms of the spherical harmonics $\mathcal{Y}_{\ell,m}$ listed in table 5.1. The $\mathcal{Y}_{\ell,m}$ are suitable for most problems of chemical bonding where visualization is important[4]. In that context, $\mathcal{Y}_{\ell,m}$ with $\ell = 0, 1, 2, 3, \ldots$ are referred to as $s, p, d, f, \ldots$ functions, respectively. Descriptive subscripts distinguish among the $\mathcal{Y}_{\ell,m}$ with the same ℓ but different m. For example, p_x, p_z, and p_y are used to denote $\mathcal{Y}_{1,m}$ with $m = -1, 0$, and $+1$ respectively. Similarly, $d_{x^2-y^2}$, d_{xz}, d_{z^2}, d_{yz}, and d_{xy} are used to denote $\mathcal{Y}_{2,m}$ with $m = +2, +1, 0, -1$ and -2, respectively. Note that the power of z in $\mathcal{Y}_{\ell,m}$ decreases as $|m|$ increases; for example, the d harmonics have z^2 when $m = 0$, z when $m = \pm 1$, and no z at all when $m = \pm 2$.

Plots of $|r^\ell \mathcal{Y}_{\ell,m}(\theta, \phi)|$ are used to visualize the wave functions. One finds a nodeless sphere for an s function, and a dumbbell-shaped object with one nodal plane for a p harmonic as shown in figure 5.5. All three p functions have the same shape but their principal axes of symmetry are oriented along the x, y, and z axes, respectively. Figure 5.5 also shows the d_{z^2} (sometimes called $d_{2z^2-x^2-y^2}$ or $d_{3z^2-r^2}$), $d_{x^2-y^2}$, and d_{xy} harmonics, each of which has two nodal planes. The principal symmetry axis of d_{z^2} is along the z axis. In the $d_{x^2-y^2}$ function, the C_2 axes through the lobes are directed along the x and y axes. On the other hand, the C_2 axes through the lobes of the d_{xy} harmonic bisect the x and y axes. The d_{yz} and d_{xz} harmonics have the same shape as the d_{xy} function but the C_2 axes through their lobes bisect the axes indicated by the subscripts.

---

[3] Review polynomials in appendix A.
[4] Choosing $\Phi_m(\phi)$ to be the complex-valued functions mentioned in section 5.1 leads to complex-valued spherical harmonics. The latter are preferred in problems involving external magnetic or electric fields in which states with different m have different energy. They are not used in this book.

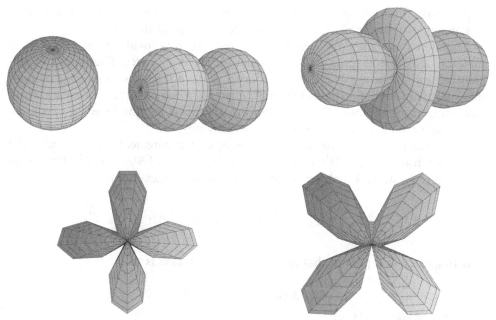

Figure 5.5. The top row shows, from left to right, the s, p_z, and d_{z^2} harmonics. The bottom row shows $d_{x^2-y^2}$ (left) and d_{xy} (right). The C_2 axes are along the x and y axes in $d_{x^2-y^2}$ but bisect them in d_{xy}.

Since they are eigenfunctions of a Hermitian operator (section 2.2), the spherical harmonics form an orthonormal set. In other words, they are normalized and orthogonal to each other:

$$\int_0^{2\pi}\int_0^{\pi} \mathcal{Y}_{L,M}(\theta,\phi)\mathcal{Y}_{\ell,m}(\theta,\phi)\sin\theta\,d\theta\,d\phi = \begin{cases} 1 & \text{if } L=\ell \text{ and } M=m \\ 0 & \text{if } L\neq\ell \text{ or } M\neq m \end{cases} \quad (5.10)$$

where the $\sin\theta$ is part of the volume element in spherical polar coordinates; see equation (5.8). This means, for example, that p type spherical harmonics are orthogonal to all harmonics of $s, d, f, \dots$ type.

5.3 The rigid rotor model

The *rigid rotor* model of a non-vibrating diatomic molecule is two atoms (balls) with masses m_1 and m_2 attached to each other by a rigid bond (rod) of length R. A center-of-mass transformation separates external from internal coordinates in the two-particle Schrödinger equation. The internal rotational motion is of interest whereas the external motion is just the translational motion of the diatomic molecule as a whole. The Schrödinger equation for the *three*-dimensional rotational motion differs from the Schrödinger equation for a particle on a sphere only in the replacement of the mass M of the particle by the reduced mass $\mu = m_1 m_2/(m_1+m_2)$ of the molecule. Hence, the spherical harmonic wave functions and the energy levels of equation (5.9), with M replaced by μ, describe the rotations of a diatomic molecule. The same energy level expression, with suitably modified definitions of I, also applies to the rotations of polyatomic linear molecules and spherical top molecules (see section 1.3). Rotations of symmetric and asymmetric tops are beyond the scope of this book.

Equation (5.9) gives the spacings between the rotational energy levels of a diatomic molecule in terms of the moment of inertia $I = \mu R^2$. Thus spectroscopically measured spacings enable us to deduce a value of I. Since we know the reduced mass μ, we can determine the bond length R from I. This constitutes an experimental measurement of the bond length. Many of the bond lengths listed in chemistry textbooks were obtained by microwave spectroscopy.

For example, a microwave spectroscopic measurement shows that the $\ell = 0 \rightarrow 1$ transition in $^2\mathrm{H}^{127}\mathrm{I}$ occurs at 196.912 GHz $= 1.96912 \times 10^{11}$ Hz. From equation (5.9), we find that the corresponding transition energy is

$$\Delta E_{0\rightarrow 1} = E_{1,m} - E_{0,m} = \frac{\hbar^2}{2I}[1(1+1) - 0(0+1)] = \frac{\hbar^2}{I} = \frac{\hbar^2}{\mu R^2}.$$

The frequency of a photon that can cause this transition is then

$$\nu = \Delta E_{0\rightarrow 1}/h = \hbar^2/(h\mu R^2).$$

The atomic masses of $m(^2\mathrm{H}) = 2.0141\ u$ and $m(^{127}\mathrm{I}) = 126.90\ u$ lead to a reduced mass $\mu = 1.9826\ u = 3.292 \times 10^{-27}$ kg. Inserting the SI values of ν, $\hbar$, h, and μ into the equation above and solving for R, we find that $R = 1.609 \times 10^{-10}$ m $= 1.609$ Å $= 160.9$ pm is the measured bond length of $^2\mathrm{H}^{127}\mathrm{I}$ (also known as DI).

More generally, a *selection rule* states that transitions between rotational levels are restricted to $\Delta\ell = \pm 1$. The energy difference for allowed transitions can be written as

$$\Delta E_{\ell\rightarrow\ell+1} = E_{\ell+1,m} - E_{\ell,m} = B[(\ell+1)(\ell+1+1) - \ell(\ell+1)]$$
$$= 2B(\ell+1) \tag{5.11}$$

in which $B = \hbar^2/(2I)$ is called the *rotational constant*. Thus the gap between measured frequencies for two adjacent transitions, $\ell \rightarrow \ell+1$ and $\ell+1 \rightarrow \ell+2$, gives $2B/h$ from which the bond length can be deduced, as in the example above.

Problems (see appendix B for hints and solutions)

5.1 List two important qualitative differences between the energy levels of a particle in a wire and those of a particle on a circular ring.

5.2 What would happen to the energy level spacing of a particle on a circular ring if the radius of the circle was doubled?

5.3 Demonstrate that the degenerate $m = \pm 1$ wave functions for a particle on a ring are normalized and orthogonal to each other.

5.4 What is the value of the following integral involving spherical harmonics $\mathcal{Y}_{\ell,m}(\theta,\phi)$?

$$\int_0^{2\pi}\int_0^{\pi}\mathcal{Y}_{2,-1}(\theta,\phi)\mathcal{Y}_{1,-1}(\theta,\phi)\sin\theta\,d\theta\,d\phi$$

5.5 Calculate the energies of the first three rotational levels of HBr using the three-dimensional rigid rotor model using $m_H = 1.008\,u$, $m_{Br} = 79.9\,u$, and $R(\text{H–Br}) = 1.414$ Å. What are the frequencies (in GHz) of the $\ell = 0 \rightarrow 1$ and $\ell = 1 \rightarrow 2$ rotational transitions?

5.6 The lowest frequency at which a rotational transition is observed experimentally in carbon monoxide is 115.271 GHz. Use the rigid rotor model to calculate the bond length of CO given the isotope-averaged atomic masses $m_C = 12.01\,u$ and $m_O = 16.00\,u$.

5.7 Explain how you might use the particle on a ring as a very simple model of the π electrons in benzene. Assume that there can be no more than two electrons per energy level on the basis of the Pauli principle that you learned in your general chemistry course. Choose R and M in a meaningful manner. Write down a formula for the wavelength of a photon that has precisely the energy required to induce the promotion of an electron from the highest filled to the lowest unfilled energy level. Do not introduce numerical values of constants into the formula but use properly defined symbols instead.

5.8
(a) Calculate $(|\mathcal{Y}_{1,-1}|^2 + |\mathcal{Y}_{1,0}|^2 + |\mathcal{Y}_{1,1}|^2)/3$.
(b) Calculate

$$(|\mathcal{Y}_{2,-2}|^2 + |\mathcal{Y}_{2,-1}|^2 + |\mathcal{Y}_{2,0}|^2 + |\mathcal{Y}_{2,1}|^2 + |\mathcal{Y}_{2,2}|^2)/5.$$

(c) Do the above results make sense as the ensemble-averaged probability densities for the $\ell = 1$ and $\ell = 2$ states of a particle on a sphere?
(d) Guess a general expression, sometimes called Unsöld's theorem, for the sum of the squared magnitudes of all the spherical harmonics with a fixed value of ℓ.

IOP Concise Physics

Quantum Chemistry
A concise introduction for students of physics, chemistry, biochemistry and materials science
Ajit J Thakkar

Chapter 6

The hydrogen atom

6.1 The Born–Oppenheimer approximation

To discover what quantum mechanics can tell us about atoms and molecules, we need to solve the Schrödinger equation $\hat{H}\psi = E\psi$. The Hamiltonian operator is $\hat{H} = \hat{T} + \hat{V}$; see equation (2.10). A molecule contains both nuclei and electrons, and so the kinetic energy operator is a sum of nuclear and electronic terms: $\hat{T} = \hat{T}_n + \hat{T}_e$. The potential energy operator $\hat{V}$ contains terms for the various Coulomb interactions between pairs of particles: electron–nucleus (en) attractions, electron–electron (ee) repulsions and internuclear (nn) repulsions. Thus $\hat{V} = V_{en} + V_{ee} + V_{nn}$ and

$$\hat{H} = \hat{T}_n + \hat{T}_e + V_{en} + V_{ee} + V_{nn}. \tag{6.1}$$

It is difficult to solve the Schrödinger equation with the Hamiltonian (6.1) and so it is common to simplify $\hat{H}$. Using the observation that nuclei move much more slowly than electrons because nuclei are much heavier than electrons[1], Max Born and J Robert Oppenheimer[2] proposed treating the electronic and nuclear motions in two separate steps.

Within the clamped nucleus or *Born–Oppenheimer approximation*, the nuclei are considered to be motionless on the timescale of electronic motion. Thus the nuclear kinetic energy $\hat{T}_n$ is omitted from $\hat{H}$ to obtain

$$\hat{H}_e = \hat{T}_e + V_{en} + V_{ee} + V_{nn}. \tag{6.2}$$

The electronic Hamiltonian $\hat{H}_e$ depends on the positions of all the particles but it contains only electronic coordinates as variables because the nuclei have fixed positions. The *electronic Schrödinger equation*

$$\hat{H}_e\psi_e = E_e\psi_e \tag{6.3}$$

---

[1] Even the lightest possible nucleus consisting of a single proton is approximately 1836 times as heavy as an electron.
[2] Oppenheimer's role in the development of nuclear weapons, his later work on nuclear non-proliferation, and his political troubles make for fascinating reading.

gives a set of electronic wave functions ψ_e and energies E_e for each fixed geometry. Observe that V_{nn} is a constant for a fixed geometry, and so

$$(\hat{H}_e - V_{nn})\psi_e = \hat{H}_e\psi_e - V_{nn}\psi_e = E_e\psi_e - V_{nn}\psi_e = (E_e - V_{nn})\psi_e. \quad (6.4)$$

Hence solving a Schrödinger equation with the Hamiltonian $\hat{H}_e - V_{nn}$ will yield the same wave functions as solving equation (6.3) and energies $(E_e - V_{nn})$ from which E_e can be obtained simply by adding the constant V_{nn}.

Repeatedly solving the electronic Schrödinger equation (6.3) at a large set of fixed nuclear geometries yields the ground-state electronic energy $E_e(\vec{R})$ as a function of the nuclear positions $\vec{R}$. The global minimum $\vec{R}_e$ of $E_e(\vec{R})$ is the *equilibrium geometry* of interest in chemistry. Other local minima on the $E_e(\vec{R})$ surface correspond to other stable isomers and first-order saddle points on the surface may correspond to transition state structures. The function $W(\vec{R}) = E_e(\vec{R}) - E_{el}$ in which $E_{el} = E_e(\vec{R}_e)$ is the *potential energy surface* upon which the nuclei move. The masses of the nuclei appear only in the nuclear kinetic energy $\hat{T}_n$. Hence the $E_e(\vec{R})$ surface and its shifted counterpart $W(\vec{R})$ are the same for all *isotopologues*— molecules that differ only in their isotopic composition (see problem 6.1).

The second and last step of the Born–Oppenheimer method is to solve the nuclear motion Schrödinger equation:

$$[\hat{T}_n + W(\vec{R})]\psi_n = E_n\psi_n. \quad (6.5)$$

The total energy of the molecule is $E = E_{el} + E_n$ and $\psi = \psi_e\psi_n$ is the overall wave function of the molecule. The solution of equation (6.5) is of central importance in molecular spectroscopy. A center-of-mass transformation followed by an approximate separation of variables (see section 4.1) allows equation (6.5) to be separated into translational (t), vibrational (v), and rotational (r) Schrödinger equations and so $E_n \approx E_t + E_v + E_r$. For a diatomic molecule, W is a function only of the bond length R and the potential energy surface is just a $W(R)$ curve of the sort seen in figure 7.5. If $W(R)$ is nearly harmonic for distances close to the equilibrium bond length R_e, then E_v can be obtained from the energy levels of a harmonic oscillator (section 3.2) and E_r from the energies of a rigid rotor (section 5.3).

Since we are concerned mostly with the electronic Schrödinger equation (6.3) in this book, we often refer to it as 'the' Schrödinger equation, and drop the e subscript on the electronic Hamiltonian and wave function.

6.2 The electronic Hamiltonian

In quantum chemistry, it is usual to work in Hartree *atomic units*. These are defined as follows. The unit of mass is the electron mass m_e, the unit of charge is the proton charge e, the unit of permittivity is the vacuum value $4\pi\varepsilon_0$, and the unit of angular momentum is $\hbar$. It then follows that the atomic unit of length is the bohr $a_0 = h^2\varepsilon_0/\pi m_e e^2$ and the atomic unit of energy is the hartree $E_h = e^2/(4\pi\varepsilon_0 a_0)$. Do not confuse the atomic unit of mass m_e with the atomic mass unit u used in general chemistry; $1\,u \approx 1836.15\,m_e$. Note that the nuclear masses m_a do not equal one in atomic units. We use atomic units almost exclusively in the rest of this book. Equations can be converted to Hartree atomic units by setting $e = \hbar = m_e = 4\pi\varepsilon_0 = a_0 = E_h = 1$. Numerical results in atomic units can be converted to other units with the conversion factors in appendix A.

We now examine, in detail, the form of the electronic Hamiltonian $\hat{H}_e$ that appears in the electronic Schrödinger equation. The total kinetic energy operator will have, for each electron, a term of the form $-(\hbar^2 \nabla_j^2)/(2m_e)$ where m_e is the mass of an electron and j is the label of the electron. In atomic units, the electronic kinetic energy of electron j becomes $-\frac{1}{2}\nabla_j^2$. The potential energy terms will contain a term for every distinct pair of particles. Recall that the Coulomb energy of interaction between a pair of particles with charges q_1 and q_2 separated by a distance r is given by $V = (q_1 q_2)/(4\pi\varepsilon_0 r)$, where ε_0 is the electric constant. In atomic units, this Coulomb energy becomes simply $V = q_1 q_2 / r$. In atomic units, the charge of an electron is -1 and the charge of a nucleus is equal to the atomic number Z. Thus, the electron–nucleus attraction V_{en} will consist of a sum of terms, one for each electron–nucleus pair, of the form $-Z_a/r_{ja}$, where r_{ja} is the distance between electron j and nucleus a with atomic number Z_a. Similarly, the electron–electron repulsion V_{ee} will be a sum of terms, one for each electron pair, of the form $1/r_{jk}$, where r_{jk} is the distance between electrons j and k. The internuclear repulsion V_{nn} will be a sum of terms, one for each nucleus pair, of the form $Z_a Z_b / R_{ab}$, where R_{ab} is the distance between nuclei a and b with atomic numbers Z_a and Z_b, respectively.

6.3 The hydrogen atom

The hydrogen atom consists of one electron and a nucleus with atomic number $Z = 1$. The nucleus is placed at the origin and the spherical polar coordinates (see figure 5.3) of the electron are (r, θ, ϕ). So r is the distance between the electron and nucleus.

$$(r, \theta, \phi)$$

$$n \diagdown \overset{r}{\diagup} \overset{e}{}$$

$$(0, 0, 0)$$

In atomic units, the electronic kinetic energy operator is $\hat{T}_e = -\nabla^2/2$ and the electron–nucleus potential is $V_{en} = -1/r$. There are no repulsions ($V_{ee} = V_{nn} = 0$) for the hydrogen atom because there is only one electron and one nucleus. Using the above operators, the electronic Schrödinger equation (6.3) is simply

$$-\frac{\nabla^2\psi}{2} - \frac{1}{r}\psi = E\psi. \tag{6.6}$$

Equation (6.6) can be solved by separation of variables in spherical polar coordinates. The angular (θ, ϕ) equation turns out to be essentially the same as that for a particle on a sphere; see section 5.2. The derivation is lengthy and only the results are summarized here.

This three-dimensional problem requires three quantum numbers: n, ℓ, and m. The boundary conditions restrict their values. The principal quantum number $n = 1, 2, \ldots$ is a positive integer, the angular momentum quantum number ℓ is restricted to $0, 1, \ldots, n-1$, and the magnetic or azimuthal quantum number m is restricted to $-\ell, -\ell+1, \ldots, \ell-1, \ell$. More concisely, $n > \ell \geqslant |m|$. As in chapter 5, the quantum numbers ℓ and m measure the state's angular momentum L: $\langle \hat{L}^2 \rangle = \ell(\ell + 1)\hbar^2$ and $\langle \hat{L}_z \rangle = m\hbar$. States with $\ell = 0, 1, 2, 3, \ldots$ are referred to as $s, p, d, f, \ldots$ states, respectively. For example, the states with $n = 2$ and $\ell = 1$ are referred to as the $2p$ states and those with $n = 3$ and $\ell = 2$ are called $3d$ states.

6.3.1 Energy levels

The spherical symmetry of the atom leads us to expect $(2\ell + 1)$-fold degeneracies like those found for a particle on a sphere. However, the energies depend only on the principal quantum number n. More precisely,

$$E_n = -\frac{1}{2n^2} E_h \qquad \text{for } n = 1, 2, \ldots \qquad (6.7)$$

in which E_h (hartree) is the atomic unit of energy. A non-geometrical symmetry beyond the scope of this book is responsible for the greater, n^2-fold, degeneracy. The $1s$ ground state is non-degenerate and it has an energy of $E_1 = -(1/2)E_h$. Degeneracies are seen in figure 6.1 for all other values of n. For example, the $2s$ state and the three $2p$ states of the hydrogen atom all have the same energy, $E_2 = -(1/8)E_h$, *unlike every other atom* where the $2s$ energy differs from the $2p$ energy. $E = 0$ is the ionization threshold above which there is a *continuum* of positive energy states. In the continuum (ionized) states, the electron is detached from the nucleus and the relative translational energy of the two particles is $E > 0$.

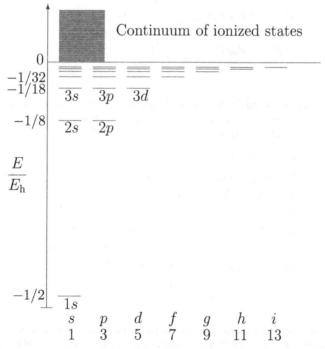

Figure 6.1. Energy levels (in atomic units) of the hydrogen atom. The degeneracy, $2\ell + 1$, of the energy levels with the same n and ℓ is shown at the bottom.

6.3.2 Orbitals

The one-electron wave functions are called *orbitals* and form an orthonormal set. They can be written as the product of a radial function, $R_{n,\ell}(r)$, and an angular function $\mathcal{Y}_{\ell,m}$. The latter is exactly a spherical harmonic (studied previously in section 5.2). Thus

$$\psi_{n,\ell,m}(r,\theta,\phi) = R_{n,\ell}(r)\mathcal{Y}_{\ell,m}(\theta,\phi). \tag{6.8}$$

The first few radial functions for the hydrogen atom are listed in table 6.1. Inserting $R_{n,\ell}(r)$ from table 6.1 and the real-valued[3] $\mathcal{Y}_{\ell,m}$ from table 5.1 into equation (6.8) gives the hydrogen orbitals listed in table 6.2. The names given to the orbitals in table 6.2 consist of the numerical value of n followed by the letter $s,p,d,f,\ldots$ corresponding to the value of ℓ and a descriptive subscript that reminds one of the angular factor for that choice of m.

Table 6.1. Radial wave functions for the hydrogen atom.

$$R_{1,0}(r) = R_{1s}(r) = 2\,e^{-r}$$

$$R_{2,0}(r) = R_{2s}(r) = \frac{1}{\sqrt{2}}(1-r/2)\,e^{-r/2}$$

$$R_{2,1}(r) = R_{2p}(r) = \frac{1}{2\sqrt{6}}r\,e^{-r/2}$$

$$R_{3,0}(r) = R_{3s}(r) = \frac{2}{9\sqrt{3}}(3-2r+2r^2/9)\,e^{-r/3}$$

$$R_{3,1}(r) = R_{3p}(r) = \frac{4}{27\sqrt{6}}(2-r/3)r\,e^{-r/3}$$

$$R_{3,2}(r) = R_{3d}(r) = \frac{4}{81\sqrt{30}}r^2\,e^{-r/3}$$

Table 6.2. Wave functions for the hydrogen atom. $N_2 = (32\pi)^{-1/2}$ and $N_3 = 2/81(2\pi)^{1/2}$ are normalization constants. Observe that all the wave functions have an $e^{-r/n}$ factor.

s	$1s = \pi^{-1/2}\,e^{-r}$	
	$2s = N_2\,(2-r)\,e^{-r/2}$	$3s = (N_3/\sqrt{6})(27-18r+2r^2)\,e^{-r/3}$
p	$2p_x = N_2\,x\,e^{-r/2}$	$3p_x = N_3\,(6-r)x\,e^{-r/3}$
	$2p_y = N_2\,y\,e^{-r/2}$	$3p_y = N_3\,(6-r)y\,e^{-r/3}$
	$2p_z = N_2\,z\,e^{-r/2}$	$3p_z = N_3\,(6-r)z\,e^{-r/3}$
d	$3d_{xy} = N_3\,xy\,e^{-r/3}$	$3d_{x^2-y^2} = (N_3/2)(x^2-y^2)\,e^{-r/3}$
	$3d_{xz} = N_3\,xz\,e^{-r/3}$	$3d_{yz} = N_3\,yz\,e^{-r/3}$
		$3d_{z^2} = (N_3/\sqrt{12})(3z^2-r^2)\,e^{-r/3}$

[3] Real-valued spherical harmonics are most convenient for the purposes of this book. Since degenerate wave functions are not unique (section 4.4), it is equally valid to use complex-valued spherical harmonics and ensure that the z component of the angular momentum is $m\hbar$ for the state $\psi_{n,\ell,m}$. This is most useful when the effects of an external electric or magnetic field are being studied.

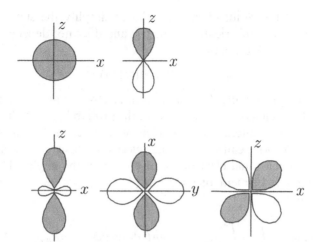

Figure 6.2. Schematic representation of the hydrogen orbitals. Top row: s and p_z orbitals. Bottom row from left to right: d_{z^2}, $d_{x^2-y^2}$ and d_{xz} orbitals.

The ground-state wave function $\psi_{1,0,0} = 1s(\vec{r}) = \pi^{-1/2}\,\mathrm{e}^{-r}$ is spherically symmetric as are all the excited s states. One way to visualize the shapes of the orbitals with $\ell > 0$ is a plot of $|r^{\ell}\mathcal{Y}_{\ell,m}|$ as in figure 5.5. Schematic representations, sufficient for most purposes, of the angular parts of the hydrogen atom orbitals are shown in figure 6.2. The convention used here is that regions where an orbital is positive are shaded and regions where it is negative are unshaded. The opposite convention is used in some books. The p_x and p_y orbitals look like the p_z orbital except that they point along the x and y axes, respectively. Schematic pictures of the d_{yz} and d_{xy} orbitals can be obtained from the picture of the d_{xz} orbital by simply relabeling the axes. These schematic pictures should be committed to memory because they are used in many qualitative arguments about molecules. The s and p orbital shapes are important in understanding the chemistry of all the elements. The d orbital shapes are important in transition metal and organometallic chemistry. Lanthanide and actinide chemists need the shapes of f orbitals.

The radial function $R_{n,\ell}$ has $n - \ell - 1$ nodes and the angular part $\mathcal{Y}_{\ell,m}$ has ℓ nodes. Hence, the overall wave function $\psi_{n,\ell,m}$ has $n - 1$ nodes. Once again, the energy increases as the number of nodes increases. The radial functions $R_{n,\ell}$ grow in size with n because the exponential factor $\mathrm{e}^{-r/n}$ decays more slowly as n increases. Since the radial functions $R_{n,\ell}$ have a r^{ℓ} factor, they vanish at the nucleus ($r = 0$) unless $\ell = 0$. It follows that only s-type radial functions ($\ell = 0$) have a non-zero value at the nucleus, an observation that has consequences for electron spin resonance spectroscopy.

6.3.3 Electron density and orbital size

The electron density ρ for a state with fixed n and ℓ is obtained by averaging over the densities for the degenerate states with the same n and ℓ but different values of m; see section 4.3. For example, we find for the $2p$ state that

$$\rho_{2p}(r, \theta, \phi) = |R_{2p}|^2 (|\mathcal{Y}_{1,-1}|^2 + |\mathcal{Y}_{1,0}|^2 + |\mathcal{Y}_{1,1}|^2)/3 = |R_{2p}|^2/(4\pi) \qquad (6.9)$$

where we have used the result of problem 5.8 to simplify the sum of the $|\mathcal{Y}_{1,m}|^2$. Notice that the density is spherically symmetric and does not depend on the angles. Similarly, for any (n, ℓ), we obtain

$$\rho_{n,\ell}(r) = |R_{n,\ell}(r)|^2/(4\pi). \tag{6.10}$$

Let us calculate the probability $D(r)\,dr$ that an electron is at a distance between r and $r + dr$ from the nucleus or, in other words, the probability of finding the electron in a thin spherical shell centered at the nucleus with inner radius r and outer radius $r + dr$. The function $D(r)$, called the radial electron density, is obtained by integrating the three-dimensional electron density over all the angles. Using equation (6.10) and the volume element of equation (5.8) appropriate for spherical polar coordinates, we find

$$D(r)\,dr = \int_0^{2\pi} \int_0^{\pi} \rho_{n,\ell}(r)\, r^2 \sin\theta\, dr\, d\theta\, d\phi = r^2 |R_{n,\ell}(r)|^2\, dr. \tag{6.11}$$

The probability that the electron will be within a sphere of radius $\mathcal{R}$ centered at the nucleus is:

$$P(\mathcal{R}) = \int_0^{\mathcal{R}} D(r)\,dr. \tag{6.12}$$

As $\mathcal{R} \to \infty$, $P(\mathcal{R}) \to 1$. The van der Waals radius of the H atom can be chosen as the value of $\mathcal{R}$ at which $P(\mathcal{R})$ reaches a number close to 1; see problem 6.11. The concept of an atom's size is fuzzy (ill-defined) because the probability densities $|\psi|^2$ and $D(r)$ do not have a finite size.

Figure 6.3 shows that the radial electron density reaches a global maximum at $r = 1$, 5.24, 4, 13.07, 12, and 9 bohrs for the $1s$, $2s$, $2p$, $3s$, $3p$, and $3d$ states,

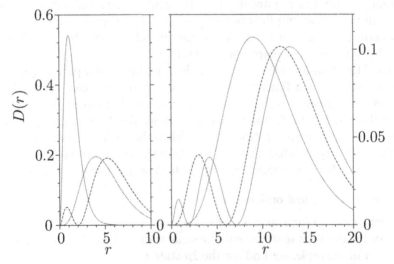

Figure 6.3. Radial electron densities for the hydrogen atom. Left panel, $1s$:——, $2s$: - - - -, and $2p$: ⋯⋯; Right panel, $3s$:——, $3p$: - - - -, and $3d$ ⋯⋯; Note the difference in the vertical scales of the two panels.

respectively. These are the most probable (or modal) electron–nucleus distances for these states. Modal electron–nucleus distances provide another measure of the size of the electron cloud associated with an orbital. We shall see their periodic behavior for atoms in section 8.5; they can be used as covalent radii. The modal distances increase rapidly with n, but differ only by small amounts for fixed n.

A third measure of size is given by the mean value of the electron–nucleus distance, which can be calculated from

$$\langle r \rangle = \int_0^\infty r\, D(r)\, \mathrm{d}r. \tag{6.13}$$

For the ground state, using equation (A.14) from appendix A, one finds:

$$\langle r \rangle = \int_0^\infty r^3 |R_{1s}(r)|^2\, \mathrm{d}r = 4 \int_0^\infty r^3\, \mathrm{e}^{-2r} \mathrm{d}r = 4 \times \frac{3!}{2^4} = \frac{3}{2} a_0$$

where a_0 (bohr) is the atomic unit of length. A more general calculation gives $\langle r \rangle = 3n^2 a_0/2$ for s states with $\ell = 0$, showing that orbital size increases quadratically with the principal quantum number n.

6.3.4 Spin angular momentum

Following Pauli's proposal of a new quantum number to explain atomic spectra, Ralph Kronig, George Uhlenbeck, and Samuel Goudsmit introduced an intrinsic angular momentum called spin for elementary particles like electrons. A simple picture of electron spin is to think of it as arising from the spinning of the electron about its own axis. Spin has to be grafted onto non-relativistic quantum mechanics but arises naturally in relativistic quantum mechanics.

In analogy with the rotational angular momentum discussed in section 5.2, the square of the spin angular momentum is $\langle \hat{S}^2 \rangle = s(s+1)\hbar^2$, where $s = \frac{1}{2}$ for electrons, protons, and neutrons. Photons have $s = 1$. Particles with half-integer spin s are called fermions after Enrico Fermi (Physics Nobel Prize, 1938), whereas those with integer spin are called bosons after Satyendra Bose. The z component of the spin angular momentum $\langle \hat{S}_z \rangle$ is restricted to the values $m_s \hbar$ with $m_s = -s, -s+1, \ldots, s-1, s$. Thus an electron has $2s + 1 = 2$ degenerate spin states, namely $m_s = \frac{1}{2}$ and $m_s = -\frac{1}{2}$, sometimes denoted $\uparrow$ and $\downarrow$, respectively. These two spin states are described by the spin functions α and β, respectively. These functions depend on a notional spin variable σ and form an orthonormal pair: $\int \alpha^2\, \mathrm{d}\sigma = 1$, $\int \beta^2\, \mathrm{d}\sigma = 1$, and $\int \alpha\beta\, \mathrm{d}\sigma = 0$.

Consideration of electron spin leads to each of the hydrogen atom orbitals $\psi_{n,\ell,m}$ being turned into two degenerate *spin orbitals* $\psi_{n,\ell,m}\alpha$ and $\psi_{n,\ell,m}\beta$. Then, the degeneracy of the hydrogen atom energy levels doubles to $2n^2$. An orbital times a spin function is called a spin orbital.

6.4 Hydrogen-like ions

The hydrogen-like ions, such as He^+ and Li^{2+}, consist of one electron and a nucleus with atomic number Z. They are no more difficult to deal with than the H atom. The relevant electronic Schrödinger equation,

$$-\frac{\nabla^2\psi}{2} - \frac{Z}{r}\psi = E\psi, \tag{6.14}$$

reduces to equation (6.6) when $Z = 1$. If we solve equation (6.14), then the solutions for H, He^+, ... can be obtained by setting $Z = 1, 2, \ldots$, respectively. The energy levels turn out to be:

$$E_{n,\ell,m} = -\frac{Z^2}{2n^2}E_h \qquad \text{for } n = 1, 2, \ldots. \tag{6.15}$$

The wave functions are again given by equation (6.8), in which the radial functions are different for different Z but the angular part is given by the spherical harmonics $\mathcal{Y}_{\ell,m}$ for all Z. The radial functions for a hydrogen-like ion with atomic number Z can be obtained from the functions in table 6.1 by multiplying them by $Z^{3/2}$ and replacing r by Zr wherever it occurs on the right-hand side. For example, from table 6.1, we find that $R_{1s}(r) = 4\sqrt{2}\,e^{-2r}$ for the helium cation He^+ which has an atomic number of $Z = 2$. The ground-state wave function for He^+, the product of R_{1s} and $\mathcal{Y}_{0,0}$, is therefore

$$1s = \sqrt{8/\pi}\,e^{-2r}. \tag{6.16}$$

The same reasoning shows that the ground-state wave function for a one-electron atom or ion with nuclear charge Z is

$$1s = \sqrt{Z^3/\pi}\,e^{-Zr}. \tag{6.17}$$

Since the decay of e^{-Zr} is faster for larger Z, the electron density becomes more compact as Z increases.

Problems (see appendix B for hints and solutions)

6.1 The vibrational force constant does not vary from one isotopologue to another; thus, for example, the force constants for H_2, HD, and D_2 are all identical. The $v = 0 \to 1$ vibrational transition in isotopically pure $^1H_2$ has been measured to occur at 4395.2 cm^{-1}. Use the harmonic oscillator model to predict the same transition in D_2.

6.2 For the hydrogen atom, specify the allowed values of ℓ for $n = 5$ and the allowed values of m for f orbitals.

6.3 Including spin considerations, give the degeneracy of the hydrogen atom energy level with (a) $n = 1$, (b) $n = 2$, and (c) $n = 3$.

6.4 Calculate, in atomic units, the energy required to excite an electron in the hydrogen atom from the ground state to the $2p_z$ state. Calculate, in nanometers, the wavelength of the photon that can cause this excitation.

6.5 Calculate, in electron volts, the energy required to ionize a hydrogen atom in the $2p_z$ state.

6.6 The mean value $\langle r \rangle$ of the distance between the nucleus and the electron in the $2p_x$ state of the hydrogen atom is $5a_0$. Without calculation, determine the mean value of the distance between the nucleus and the electron in the $2p_z$ state of the hydrogen atom. Explain your reasoning.

6.7 Calculate the expectation value $\langle r \rangle$ for the $3p$ and $3d$ states of the hydrogen atom. Use either integral formulas from appendix A or mathematical software.

6.8 Calculate the expectation value $\langle r \rangle$ for the ground state of the He^+ cation. Compare your result with $\langle r \rangle$ for the H atom.

6.9 Write down, in atomic units, the $2s$ orbital for He^+.

6.10 Without calculation, write down the value of the following integral involving hydrogen atom orbitals. Explain your answer.

$$\int_0^{2\pi} \int_0^{\pi} \int_0^{\infty} 3d_{xz}(\vec{r})\, 2p_z(\vec{r})\, r^2 \sin\theta \, dr \, d\theta \, d\phi$$

6.11
(a) Derive $P(\mathcal{R})$, the probability that an electron will be no further than a distance $\mathcal{R}$ from the nucleus, for the ground state of the hydrogen atom. Use equation (6.12) and either integral formulas from appendix A or mathematical software.
(b) Plot the above $P(\mathcal{R})$ and use it to find the electron–nucleus distances $\mathcal{R}$ at which the probability $P(\mathcal{R})$ becomes 0.85, 0.95, and 0.99.
(c) Compare these distances with the van der Waals radius of H from an inorganic chemistry textbook. Which of these distances best matches the van der Waals radius? What is the probability of finding the electron further from the nucleus than this radius? Does the answer to the preceding question make you uncomfortable? Explain why or why not.

Quantum Chemistry
A concise introduction for students of physics, chemistry, biochemistry and materials science
Ajit J Thakkar

Chapter 7

A one-electron molecule: H_2^+

The simplest molecule is the hydrogen molecule ion H_2^+. It consists of two hydrogen nuclei (protons) and one electron. The nuclei are stationary within the Born–Oppenheimer approximation, and they are taken to be a fixed distance R apart. The z axis is chosen to lie along the principal axis of symmetry, and the origin is placed at the center of mass. As depicted in figure 7.1, the Cartesian coordinates of the electron are (x_1, y_1, z_1), the coordinates of the two nuclei a and b are $(0, 0, -R/2)$ and $(0, 0, +R/2)$, respectively, and r_{1a} and r_{1b} are the distances between the electron and the protons.

Using the ideas of section 6.2, the electronic Schrödinger equation is:

$$\hat{H}_e \psi_e = \left(-\frac{1}{2}\nabla_1^2 - \frac{1}{r_{1a}} - \frac{1}{r_{1b}} + \frac{1}{R} \right) \psi_e = E_e \psi_e. \tag{7.1}$$

The first term in $\hat{H}_e$ is the electronic kinetic energy $\hat{T}_e$ and the next two terms make up the electron–nuclear attraction V_{en}. The internuclear repulsion $1/R$ is a constant because R is fixed. The interelectronic repulsion $V_{ee} = 0$ because there is only one electron. Since H_2^+ has only one electron, the wave function will be a *molecular orbital* (MO).

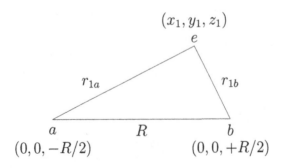

Figure 7.1. Coordinates for H_2^+.

doi:10.1088/978-1-627-05416-4ch7

Equation (7.1) can be solved exactly after transformation to confocal elliptic coordinates: $\mu = (r_{1a} + r_{1b})/R$, $\nu = (r_{1a} - r_{1b})/R$, and the azimuthal angle ϕ. The exact wave functions can be written as $\psi(\mu, \nu, \phi) = F(\mu)\, G(\nu)\, \Phi_m(\phi)$ where $\Phi_m(\phi)$ is a particle-on-a-ring wave function from equation (5.4). The wave functions with $m = 0, \pm 1, \pm 2, \ldots$ are called $\sigma, \pi, \delta, \ldots$ states, respectively. (Read δ as dell-tah.) Since there is a center of symmetry, the state label is given a subscript of either g (for *gerade*) or u (for *ungerade*), depending on whether the wave function is symmetric or antisymmetric with respect to inversion. Unfortunately, the exact electronic wave functions are rather complicated. For our purposes, it is sufficient to examine simple approximate wave functions.

7.1 The LCAO model

If the protons are far apart and the electron is close to one of them, say nucleus a, then the MO should be approximately the $1s$ orbital of an isolated hydrogen atom located at a. Hence, a reasonable trial MO can be written as a linear combination of atomic orbitals (LCAO):

$$\psi = c_a 1s_a + c_b 1s_b \tag{7.2}$$

where $1s_a = \pi^{-1/2}\, e^{-r_{1a}}$ and $1s_b = \pi^{-1/2}\, e^{-r_{1b}}$, respectively, are $1s$ atomic orbitals (AOs) on protons a and b, and c_a and c_b are mixing coefficients. We also say that the MO has been expanded in terms of a minimal *basis set* of one $1s$ orbital on each atom.

The electron probability density ρ is

$$\rho(\vec{r}) = \psi^2 = c_a^2 (1s_a)^2 + c_b^2 (1s_b)^2 + 2 c_a c_b 1s_a 1s_b. \tag{7.3}$$

The electron density ρ must be inversion symmetric because the molecule has $D_{\infty h}$ symmetry, and hence $c_b^2 = c_a^2$. This implies that $c_b = \pm c_a$. Thus, there are two MOs given by

$$1\sigma_g = N_g(1s_a + 1s_b) \qquad \text{and} \qquad 1\sigma_u = N_u(1s_a - 1s_b) \tag{7.4}$$

where N_g and N_u are normalization constants. The names chosen for the MOs are based on symmetry. The labeling convention for the m quantum number tells us that both MOs represent σ states (read σ as sigma) because they are combinations of AOs with $m = 0$. The $1\sigma_g$ and $1\sigma_u$ MOs have a g and u designation, respectively, since they are symmetric and antisymmetric with respect to inversion. Anticipating that they are the lowest-energy σ states, we call them the $1\sigma_g$ and $1\sigma_u$ MOs. To obtain the complete electronic wave functions, the MOs are multiplied by a spin function (either α or β). All terms in equation (7.3) are positive when $c_b = c_a$, so the $1\sigma_g$ MO leads to a build-up of electron density between the protons and is therefore *bonding* in nature. Letting $c_b = -c_a$ makes the last term in equation (7.3) negative. This shows that the $1\sigma_u$ MO leads to a diminishing of electron density between the protons because it has a nodal plane halfway between the nuclei and perpendicular to the molecular axis. The $1\sigma_u$ MO is therefore *antibonding* and higher in energy than the $1\sigma_g$ MO.

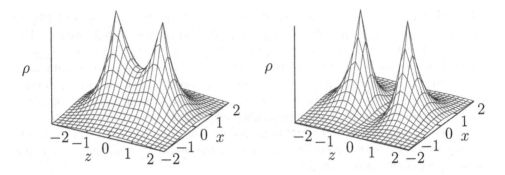

Figure 7.2. Perspective plots of the electron density $\rho(\vec{r})$ in the xz plane for the bonding $1\sigma_g$ (left) and anti-bonding $1\sigma_u$ (right) states of H_2^+.

Figure 7.2 shows perspective plots of the electron density ρ in the xz plane. Clearly the electron density peaks at the nuclei. In the $1\sigma_g$ state there is a build-up of density between the nuclei, and there is a saddle point at the middle of the bond. Such a saddle in the electron density is characteristic of a chemical bond. In general, the extrema (maxima, minima and saddle points) of the electron density have chemical significance. In the $1\sigma_u$ state, we see a depletion of density and a nodal plane between the nuclei.

7.2 LCAO potential energy curves

We now turn to the calculation of energies for the ground state of H_2^+. The $1\sigma_g$ wave function (7.4) can be normalized by choosing N_g appropriately using the method shown on page 2-5. First calculate the normalization integral.

$$\int 1\sigma_g^2 \, d\tau = N_g^2 \left(\int 1s_a^2 \, d\tau + \int 1s_b^2 \, d\tau + 2 \int 1s_a 1s_b \, d\tau \right)$$

$$= N_g^2 (2 + 2S_{ab}) \tag{7.5}$$

because the $1s$ AOs are normalized; $S_{ab} = \int 1s_a 1s_b \, d\tau$ is called an *overlap integral*. The N_g normalization constant is now determined by setting the right-hand side of equation (7.5) equal to one; thus

$$N_g = (2 + 2S_{ab})^{-1/2}. \tag{7.6}$$

The energy for the $1\sigma_g$ wave function is found by calculating the expectation value of the energy (Hamiltonian) operator using equation (2.17). Hence

$$E_g = \int 1\sigma_g (\hat{H}_e 1\sigma_g) \, d\tau = N_g^2 (H_{aa} + H_{ab} + H_{ba} + H_{bb}) \tag{7.7}$$

where

$$H_{pq} = \int 1s_p (\hat{H}_e 1s_q) \, d\tau \qquad \text{for } p, q = a \text{ or } b. \tag{7.8}$$

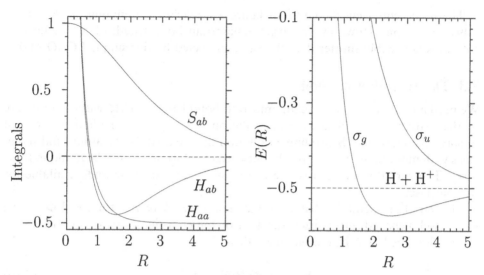

Figure 7.3. The LCAO model of H_2^+. Left panel: integrals. Right panel: potential energy curves for the σ_g and σ_u states.

We see that $H_{bb} = H_{aa}$ by symmetry. Moreover, $H_{ba} = H_{ab}$ because $\hat{H}_e$ is both real and Hermitian (see equation (2.6)) and the $1s$ AOs are real-valued. Thus, we find

$$E_g = 2N_g^2(H_{aa} + H_{ab}) = \frac{H_{aa} + H_{ab}}{1 + S_{ab}} \tag{7.9}$$

where equation (7.6) is used to obtain the final expression. A similar calculation gives $N_u = (2 - 2S_{ab})^{-1/2}$ and $E_u = (H_{aa} - H_{ab})/(1 - S_{ab})$ for the $1\sigma_u$ state.

Note from figure 7.3 how the overlap integral S_{ab} varies with the bond length R. As $R \to 0$, the two AOs become the same and $S_{ab} \to 1$ because the AOs are normalized. As R gets larger, the overlap decreases; hence $S_{ab} \to 0$ as $R \to \infty$. The same behavior would be obtained for the overlap of any two equivalent normalized AOs placed on equivalent nuclei a and b.

Figure 7.3 shows that as R becomes large, $H_{aa} = \int 1s_a(\hat{H}_e 1s_a)\,d\tau$ correctly approaches the energy of an isolated hydrogen atom, $-0.5\ E_h$. H_{ab} is a measure of the interaction between the $1s_a$ and $1s_b$ functions. Observe in figure 7.3 that, as the two orbitals approach, H_{ab} becomes increasingly attractive until the overlap becomes sizeable. Then, as the bond length is shortened further, H_{ab} rapidly becomes less attractive and becomes repulsive.

Use of H_{aa}, H_{ab} and S_{ab} in equation (7.9) and its σ_u analogue leads to potential energy curves for the σ_g and σ_u MOs, respectively. Figure 7.3 shows that, for all bond lengths $R > 1.6\ a_0$, the σ_g potential energy curve is lower than $-0.5\ E_h$, the sum of the energy of an isolated hydrogen atom and a bare proton. In other words, the ground $1\sigma_g$ state of H_2^+ is predicted to be stable. On the other hand, the σ_u potential energy curve is higher in energy than the dissociation products for all bond lengths and corresponds to an unbound excited state.

The σ_g potential energy curve in figure 7.3 reaches a minimum at $R = R_e = 2.49\,a_0 = 132$ pm. However, the exact equilibrium bond length of $R_e = 2.00\,a_0 = 106$ pm is noticeably shorter than the value predicted by this simple LCAO-MO.

7.3 The variation method

We need a method to calculate an improved bond length for H_2^+ without resorting to the extremely complicated exact solution. Moreover, a method is required to solve the electronic Schrödinger equation approximately for atoms and molecules with more than one electron because exact solutions cannot be found in these cases. The variation method is one of the most important methods available for this purpose.

The variation method is based on the *variational theorem*: If Φ (read Φ as fye) is any well-behaved, normalized function of the coordinates of the particles of a system described by a Hamiltonian $\hat{H}$, then

$$E_\Phi = \int \Phi(\hat{H}\Phi)\,\mathrm{d}\tau \geqslant E_{gs} \qquad (7.10)$$

where E_{gs} is the true ground-state energy of the system. A well-behaved function is one that is continuous, square-integrable, satisfies the boundary conditions of the problem, and, as we shall see in section 8.2, obeys the Pauli postulate if the system contains two or more identical particles. Since the integral in equation (7.10) is just the expectation value of the energy operator using the approximate wave function Φ, the inequality in the variational theorem becomes an equality if $\Phi = \psi_{gs}$, where ψ_{gs} is the true ground-state wave function. The non-trivial content of equation (7.10) is that forming an energy expectation value with an approximate wave function leads to an *upper* bound to the true ground-state energy[1]. If one prefers to work with unnormalized trial functions Φ, then equation (7.10) can be rewritten as

$$E_\Phi = \frac{\int \Phi(\hat{H}\Phi)\,\mathrm{d}\tau}{\int |\Phi|^2\,\mathrm{d}\tau} \geqslant E_{gs}. \qquad (7.11)$$

The ratio of integrals in equation (7.11) is often called the Rayleigh–Ritz quotient. If Φ is complex-valued, then the leftmost Φ in the integrals in equations (7.10)–(7.11) must be replaced by Φ^*.

A simple application of equation (7.11) is to calculate the ground-state energy predicted by a guessed wave function as in section 7.2; figure 7.4 confirms that, as predicted by the variational theorem, the LCAO energy curve lies above the exact one. In realistic applications, a trial function contains parameters that are

---

[1] This happens because it is always possible to express Φ as a linear combination, with coefficients c_k, of *all* the eigenfunctions ψ_k of $\hat{H}$. The energy of Φ is then a linear combination of the corresponding energies E_k with squared coefficients c_k^2. The latter combination of energies is greater than E_{gs} because all the $c_k^2 > 0$ and all the $E_k \geqslant E_{gs}$.

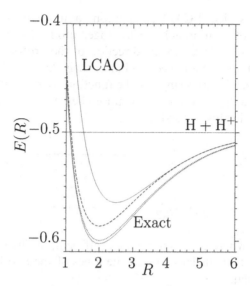

Figure 7.4. Potential energy curves for the ground state of H_2^+. The simple LCAO and exact curves are shown as solid lines; sandwiched in between are the curves for the scaled LCAO (- - - -) and the polarized model ($\cdots\cdots$).

systematically varied to determine the values that lead to the lowest E_Φ. Then, the trial function Φ that gives the lowest E_Φ is selected as the best approximation to the true ground-state wave function and E_Φ as the best approximation to the ground-state energy. The variational method is used in this way in section 7.4 to improve the LCAO wave function for H_2^+.

7.4 Beyond the LCAO model

The main deficiency of the simple LCAO model of section 7.2 is that it incorporates only large R behavior. In the united atom limit, $R \to 0$, $H_2^+ \to He^+$, and the $1s$ orbital should behave as e^{-2r}; see equation (6.16). Moreover, the size of the atomic orbitals should be allowed to adapt to the molecular environment at intermediate values of R. This behavior can be incorporated in the MO of equation (7.4) by using hydrogen-like ion AOs of the form (see equation (6.17)):

$$1s_a = \sqrt{\zeta^3/\pi}\, e^{-\zeta r_{1a}} \quad \text{and} \quad 1s_b = \sqrt{\zeta^3/\pi}\, e^{-\zeta r_{1b}} \tag{7.12}$$

where ζ (read ζ as zay-tah) is an *effective nuclear charge* which changes as the bond length R changes. The integrals S_{ab}, H_{aa} and H_{ab}, and the energy are functions of both ζ and R. The variational principle (7.10) tells us that, at each fixed R, we should choose the ζ that leads to the lowest energy making sure to force $\zeta > 0$ to keep the approximate wave function square-integrable. Doing this calculation gives ζ values changing smoothly from $\zeta = 2$ at $R = 0$ to $\zeta = 1$ as $R \to \infty$. The resulting $E(R)$ shown in figure 7.4 has a minimum at $R = R_e = 107$ pm in good agreement with the experimental $R_e = 106$ pm.

To improve the scaled LCAO calculation, a more flexible trial function is required. When a proton approaches a hydrogen atom, the electron density of the hydrogen atom should polarize in the direction of the proton. In other words, the atomic orbitals must also be allowed to change their shape in the molecule. We can incorporate this effect by allowing the $1s$ function to be polarized by mixing or 'hybridizing' it with a $2p_z$ function as illustrated below. The $2p_z$ function is called a *polarization function* in this context.

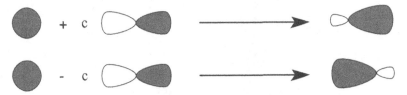

We use $1s_a + c\,2p_{za}$, where $c > 0$ is a mixing coefficient, on the left-hand nucleus and $1s_b - c\,2p_{zb}$ on the other nucleus so that the lobes of the two hybridized functions have a positive overlap.

Thus, the polarized wave function for the σ_g state is

$$\psi = N[(1s_a + c\,2p_{za}) + (1s_b - c\,2p_{zb})] \qquad (7.13)$$

in which we use hydrogenic $1s$ functions as in equation (7.12), N is a normalization constant, and the hydrogenic $2p_z$ functions are given by

$$2p_{za} = \sqrt{\beta^5/\pi}\, z_{1a}\, e^{-\beta r_{1a}} \qquad \text{and} \qquad 2p_{zb} = \sqrt{\beta^5/\pi}\, z_{1b}\, e^{-\beta r_{1b}} \qquad (7.14)$$

where the exponent β is another parameter. The MO (7.13) is said to be expanded in a *polarized* basis set consisting of four functions: $1s_a$, $1s_b$, $2p_{za}$, and $2p_{zb}$. The ideas of polarization functions and polarized basis sets are of general utility and will recur in section 10.2.

Minimization of the energy with respect to the three parameters, ζ, c, and β, for a series of bond lengths R leads to the much improved potential energy curve shown in figure 7.4. This curve has a minimum at $R = R_e = 106$ pm in perfect agreement with the exact value. At the equilibrium bond length R_e, the optimal parameters for the polarized LCAO wave function are $\zeta = 1.246$, $\beta = 1.482$, and $c = 0.138$. Observe that the optimal ζ and β indicate that the $1s$ and $2p_z$ functions are of approximately the same size as they should be for $2p_z$ to polarize the $1s$ function. We now turn to the evaluation of two other important properties of H_2^+—the dissociation energy and vibrational force constant.

7.5 Force constant and dissociation energy

To predict the rotational and vibrational energy levels of a diatomic molecule such as H_2^+, we should insert $W(R) = E(R) - E(R_e)$ into the Schrödinger equation for nuclear motion, equation (6.5), and solve by separation of variables. Since full detail is not needed, we use the approximate procedure described in section 6.1. Make a

Taylor expansion of $E(R)$ around the equilibrium bond length R_e, with $x = R - R_e$ as the expansion variable, as follows:

$$E(R) = E(R_e) + \left(\frac{dE}{dR}\right)_{R_e} x + \frac{1}{2}\left(\frac{d^2E}{dR^2}\right)_{R_e} x^2 + \cdots \quad (7.15)$$

Note that dE/dR vanishes at $R = R_e$ because $E(R)$ has a minimum there, neglect x^j terms with $j \geqslant 3$, and define the harmonic *force constant* $k_e = (d^2E/dR^2)_{R_e}$. Then $W(R) \approx k_e x^2/2$ which is a harmonic potential. So we use section 3.2 to find that the oscillator's vibrational energy levels, in atomic units, are $E_v = \omega(v + 1/2)$ in which $\omega = (k_e/\mu)^{1/2}$, μ is the reduced mass of the molecule, and v is the vibrational quantum number. Then, the zero-point ($v = 0$) vibrational energy is $\omega/2$. The harmonic energy-level spacing will only be accurate for low-lying vibrational levels because $E(R)$ is harmonic only for R close to R_e. Application of this procedure to the potential energy curve predicted by the wave function (7.13) leads to a predicted $\omega = 2315 \text{ cm}^{-1}$ to be compared with the exact value of 2297 cm^{-1} for H_2^+. As mentioned in section 6.1, rotational energies can be computed from the rigid rotor expression (5.9) with $I = \mu R_e^2$.

The *equilibrium dissociation energy* D_e of a diatomic molecule is the energy required to separate the diatomic at its equilibrium bond length R_e into atomic fragments; see figure 7.5. Using the energy predicted by the wave function of equation (7.13) we find, in atomic units, for H_2^+ that

$$D_e = E(H) + E(H^+) - E(H_2^+, R_e) = -0.5 + 0.0 - (-0.6004) = 0.1004 \, E_h.$$

Using a conversion factor from appendix A, we find $D_e = 263 \text{ kJ mol}^{-1}$ to be compared with the exact value of 269 kJ mol^{-1}.

Since a molecule is always vibrating, the dissociation energy relative to the ground vibrational state, $D_0 = D_e - \omega/2$ (see figure 7.5), is often more relevant than D_e.

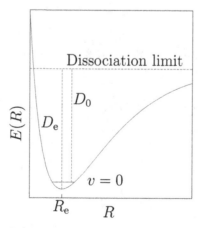

Figure 7.5. The equilibrium dissociation energy D_e and the dissociation energy D_0 from the vibrational ground state $v = 0$.

If the molecular gas can be well approximated as a perfect gas, then D_0 is equal to the enthalpy of reaction for the process: diatomic(g) $\rightarrow$ atoms(g) at $0\,\mathrm{K}$. D_0 is also called the *atomization energy*. The four quantities R_e, D_e, D_0, and k_e are all accessible through experiment.

7.6 Excited states

Figure 7.6 shows excited state MOs for H_2^+ made from higher-energy AOs of H. The π labels follow from the $m = \pm 1$ quantum numbers of the AOs involved (see table 5.1) and the naming scheme given on page 7-2. The σ_g and π_u MOs are bonding MOs, whereas the σ_u and π_g MOs are antibonding MOs. Formation of bonding MOs requires constructive overlap of the AOs whereas destructive overlap of AOs leads to antibonding MOs. Observe that the more nodal planes an MO has, the greater its energy.

The MOs of H_2^+ are used as prototypes for the MOs of A_2 diatomic molecules, and so it is important to be able to sketch them. The MOs with the same symmetry label such as σ_g are sequentially numbered in order of increasing energy. In some inorganic chemistry texts, the core 1σ MOs are dropped and the remaining σ MOs are renumbered starting from 1.

Mixing of the $2s$ and $2p_z$ AOs is expected in the 2σ and 3σ MOs because of the degeneracy of $2s$ and $2p$ AOs in the H atom. However, it has been ignored here

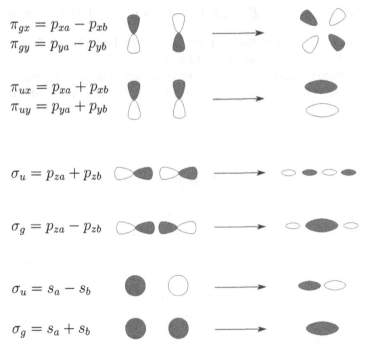

$$\pi_{gx} = p_{xa} - p_{xb}$$
$$\pi_{gy} = p_{ya} - p_{yb}$$

$$\pi_{ux} = p_{xa} + p_{xb}$$
$$\pi_{uy} = p_{ya} + p_{yb}$$

$$\sigma_u = p_{za} + p_{zb}$$

$$\sigma_g = p_{za} - p_{zb}$$

$$\sigma_u = s_a - s_b$$

$$\sigma_g = s_a + s_b$$

Figure 7.6. Formation of typical diatomic MOs from AOs. The shaded lobes are positive and the unshaded ones negative. The π_{ux} and π_{gx} MOs are in the xz plane and the π_{uy} and π_{gy} MOs are in the yz plane.

because these MOs are meant for a qualitative description of all homonuclear diatomic molecules A_2, and $2s$ AOs are not degenerate with $2p$ AOs in many-electron atoms. The energy ordering in H_2^+ is

$$1\sigma_g < 1\sigma_u < 2\sigma_g < 2\sigma_u < 1\pi_{ux} = 1\pi_{uy} < 3\sigma_g < 1\pi_{gx} = 1\pi_{gy} < 3\sigma_u. \qquad (7.16)$$

The ordering varies a bit in other A_2 molecules.

Problems (see appendix B for hints and solutions)

7.1 Write down, in atomic units, the electronic Hamiltonian for the one-electron molecule HeH^{2+}. Explain the meaning of all symbols.

7.2 Prove that the $1\sigma_g$ and $1\sigma_u$ MOs of equation (7.4) are orthogonal.

7.3 Construct electron configurations for H_2 and He_2 using the $1\sigma_g$ and $1\sigma_u$ MOs. Can you see how the electron configuration indicates that He_2 is not a stable molecule?

7.4 Tony used the variational method for the hydrogen atom with $(\zeta^3/\pi)^{1/2} e^{-\zeta r}$ as a trial function and varied ζ to obtain the lowest possible energy. What value of ζ and the energy should he have found and why? No calculations are required.

7.5 Suppose that overlap is neglected by assuming that $S_{ab} = 0$. How does this change the energies of the LCAO σ_g and σ_u wave functions?

7.6 As a simple example of the use of the variation method, suppose that we did not know how to solve the problem of a particle moving freely in a one-dimensional wire with infinite walls. Consider

$$\Phi(x) = \begin{cases} x(a-x) & \text{for } 0 \leqslant x \leqslant a, \\ 0 & \text{for } x \geqslant a \text{ and } x \leqslant 0 \end{cases}$$

as a trial function. It is well-behaved, it is continuous, square-integrable, and it vanishes at the walls.
(a) Calculate $\int \Phi(\hat{H}\Phi)\, dx$ where $\hat{H}$ is the Hamiltonian for a particle in a one-dimensional wire.
(b) Calculate $\int |\Phi|^2\, dx$.
(c) Use the results obtained above in equation (7.11) to calculate the ground-state energy predicted by this simple trial function. Compare it with the exact ground-state energy for this problem.

7.7 Write and sketch two MOs made up of d_{xz} AOs placed on equivalent nuclei a and b. Assign them appropriate symbols (such as σ_g or ϕ_u) explaining your choices. Which MO is bonding and which antibonding? Are either or both of them degenerate?

7.8 Write and sketch two MOs made up of d_{xy} AOs placed on equivalent nuclei a and b. Assign them appropriate symbols (such as σ_g or ϕ_u), explaining your choices. Which MO is bonding and which antibonding? Are either or both of them degenerate?

7.9 Isotopically pure $^1\text{H}_2^+$ has an equilibrium bond length $R_e = 106\,\text{pm}$, an equilibrium dissociation energy $D_e = 269\,\text{kJ mol}^{-1}$, and the fundamental vibrational frequency $\omega = 2297\,\text{cm}^{-1}$. The atomic masses of ^1H and ^2H (also known as D) are 1.0078 and $2.014\,u$, respectively.

(a) Calculate the atomization energy D_0 for $^1\text{H}_2^+$.

(b) What is the value of R_e for D_2^+? Explain your answer.

(c) What is the value of D_e for D_2^+? Explain your answer.

(d) Calculate the vibrational frequency ω for D_2^+.

(e) Calculate the atomization energy D_0 for D_2^+.

Quantum Chemistry

A concise introduction for students of physics, chemistry, biochemistry and materials science

Ajit J Thakkar

Chapter 8

Many-electron systems

8.1 The helium atom

The helium atom consists of two electrons and a He^{2+} nucleus (α particle) of charge +2, as shown below. r_1 and r_2, respectively, are the distances between electrons 1 and 2 and the nucleus, and r_{12} is the interelectronic distance.

Following section 6.2, the electronic Hamiltonian is

$$\hat{H}_e = -\frac{1}{2}\nabla_1^2 - \frac{1}{2}\nabla_2^2 - \frac{2}{r_1} - \frac{2}{r_2} + \frac{1}{r_{12}}. \tag{8.1}$$

The first two terms make up the electronic kinetic energy and the next two the electron–nucleus attraction. The interelectronic repulsion term, $1/r_{12} = 1/[(x_1 - x_2)^2 + (y_1 - y_2)^2 + (z_1 - z_2)^2]^{1/2}$, prevents use of the separation of variables technique of section 4.1. Therefore, *exact solutions for He or any other system with two or more electrons cannot be obtained; finding approximate ones is the central task of quantum chemistry.*

Let us begin with an *approximate* factorization of the wave function into two orbitals, one for each electron, as follows:

$$\psi(\vec{r}_1, \vec{r}_2) = 1s(\vec{r}_1)1s(\vec{r}_2). \tag{8.2}$$

First, we try using $1s = \sqrt{8/\pi}\, e^{-2r}$, the $1s$ orbital for He^+ from section 6.4, because it should be similar to the $1s$ orbital in He. With this guess in equation (8.2), we find from equation (7.11) that $E = -2.75\,E_{\mathrm{h}}$. This predicts an ionization energy $I = E(\mathrm{He^+}) - E(\mathrm{He}) \approx -2\,E_{\mathrm{h}} - (-2.75\,E_{\mathrm{h}}) = 0.75\,E_{\mathrm{h}} = 1970\,\mathrm{kJ\,mol^{-1}}$ which is far from the experimental value of $I = 2373\,\mathrm{kJ\,mol^{-1}}$. A better choice for the orbital is to use a $1s$ orbital appropriate for a hydrogen-like ion with an effective nuclear charge ζ,

$$1s(\vec{r}) = (\zeta^3/\pi)^{1/2}\,\mathrm{e}^{-\zeta r}, \tag{8.3}$$

and determine ζ by the variational method (see section 7.3). Thus the model wave function becomes $\psi(\vec{r}_1, \vec{r}_2) = (\zeta^3/\pi)\,\mathrm{e}^{-\zeta r_1}\,\mathrm{e}^{-\zeta r_2}$. Substituting it into equation (7.11) and integrating gives $E = \zeta^2 - 27\zeta/8$. Then we minimize E with respect to ζ by setting $\mathrm{d}E/\mathrm{d}\zeta = 0$. This leads to $\zeta = 27/16 = 1.6875$, $E = -2.8477\,E_{\mathrm{h}}$, and a significantly improved value of $I = 2226\,\mathrm{kJ\,mol^{-1}}$. The effective nuclear charge is smaller than the atomic number ($Z = 2$) because one electron is *shielded* from the nucleus by the other electron.

8.2 Spin and the Pauli postulate

In the hydrogen atom, we simply multiplied the $1s$ orbital by either α or β to obtain a spin orbital. In a similar way, we multiply the wave function (8.2) by an α for electron 1 and a β for electron 2 to obtain:

$$\psi(\vec{r}_1, \sigma_1, \vec{r}_2, \sigma_2) = 1s(\vec{r}_1)\,1s(\vec{r}_2)\alpha(\sigma_1)\beta(\sigma_2). \tag{8.4}$$

To simplify the notation, let us use k to stand for $\vec{r}_k, \sigma_k$ so that, for example, $\psi(1, 2) = \psi(\vec{r}_1, \sigma_1, \vec{r}_2, \sigma_2)$. Further, instead of showing the α and β spins explicitly, we use a bar to indicate those spin orbitals that have β spin. For example, $1s(k) = 1s(\vec{r}_k)\alpha(\sigma_k)$ and $\overline{1s}(k) = 1s(\vec{r}_k)\beta(\sigma_k)$. Now we can write equation (8.4) as

$$\psi(1, 2) = 1s(1)\overline{1s}(2). \tag{8.5}$$

Is this correct? Wolfgang Pauli provided the answer as an additional postulate of quantum mechanics.

Postulate 6 *The wave function of a system containing two or more electrons changes sign if the space and spin coordinates of any pair of electrons are interchanged.*

Wave functions which satisfy the Pauli condition are said to be antisymmetric with respect to interchange of electrons. The Pauli postulate is applicable to all fermions (see section 6.3.4); a general form of the postulate requires wave functions for systems containing two or more identical bosons to remain unchanged when the space and spin coordinates of any pair of identical bosons are interchanged.

For a two-electron system, the Pauli postulate requires that the wave function be antisymmetric with respect to simultaneous interchange of electron coordinates and spins:

$$\psi(1, 2) = -\psi(2, 1). \tag{8.6}$$

However, the wave function in equation (8.5) does not satisfy the Pauli requirement of equation (8.6) because interchanging the space and spin coordinates of the two electrons leads to

$$-\psi(2,1) = -1s(2)\overline{1s}(1) = -\overline{1s}(1)1s(2), \qquad (8.7)$$

which is clearly not equal to equation (8.5). A little thought shows that a combination of the right-hand sides of equation (8.5) and equation (8.7) given by

$$\psi(1,2) = [1s(1)\overline{1s}(2) - \overline{1s}(1)1s(2)]/\sqrt{2} \qquad (8.8)$$

does satisfy the Pauli condition of equation (8.6). The factor of $1/\sqrt{2}$ keeps the wave function normalized. Equation (8.8) is the product of wave function (8.2) and a spin part:

$$[\alpha(\sigma_1)\beta(\sigma_2) - \beta(\sigma_1)\alpha(\sigma_2)]/\sqrt{2}.$$

Thus spin can be tacked on at the end for a two-electron system in just a slightly more complicated way than for one-electron systems, and all the results of section 8.1 are valid.

A systematic procedure for generating antisymmetric wave functions was found by John Slater. He noticed that the wave function of equation (8.8) can be rewritten as a 2×2 determinant[1]:

$$\psi(1,2) = \frac{1}{\sqrt{2!}} \det \begin{vmatrix} 1s(1) & 1s(2) \\ \overline{1s}(1) & \overline{1s}(2) \end{vmatrix}. \qquad (8.9)$$

Slater saw that the determinant (8.9) can be generalized for many-electron atoms or molecules to provide a wave function that satisfies the Pauli condition. For example, a 3×3 Slater determinant for the ground state of the lithium atom can be written as:

$$\psi(1,2,3) = \frac{1}{\sqrt{3!}} \det \begin{vmatrix} 1s(1) & 1s(2) & 1s(3) \\ \overline{1s}(1) & \overline{1s}(2) & \overline{1s}(3) \\ 2s(1) & 2s(2) & 2s(3) \end{vmatrix}. \qquad (8.10)$$

The rows correspond to spin orbitals and the columns to electrons. In an even more compact notation, a Slater determinant is specified by a list of the spin orbitals it is constructed from. For example, the Slater determinants of equation (8.9) and equation (8.10) can be abbreviated as $|1s\overline{1s}|$ and $|1s\overline{1s}2s|$, respectively.

Interchanging all the coordinates of a pair of electrons amounts to interchanging a pair of columns in the Slater determinant. But determinants change sign when a pair of their columns (or rows) is interchanged. Thus we see that the Slater determinant ensures that the Pauli postulate is satisfied. Putting two electrons into the same spin orbital makes two rows of the Slater determinant identical. But determinants

_____
[1] Appendix A reviews determinants.

with two identical rows (or columns) are equal to zero. This shows why *there can be no more than one electron per spin orbital and no more than two electrons per orbital*. The latter statement is precisely the Pauli principle taught in introductory chemistry courses.

8.3 Electron densities

Postulate 1 on page 2-2 tells us that, for a two-electron atom, $|\psi|^2 \, d\vec{r}_1 \, d\vec{r}_2$ is the probability of finding an electron within an infinitesimal volume centered at $\vec{r}_1$ and the other electron in an infinitesimal volume centered at $\vec{r}_2$. $|\psi|^2$ is a complicated function of six variables even though we have ignored spin. The situation gets much worse for larger systems with N_e electrons because $|\psi|^2$ is then a function of $3N_e$ space variables and N_e spin variables. Such a probability distribution contains far more information than we can digest. For many chemical problems, all we need is the *electron density* $\rho(\vec{r})$, the probability of finding an electron at a point in space irrespective of where the remaining electrons in the system are. ρ can be calculated from the many-electron probability density by averaging over the positions of the other $N_e - 1$ electrons and the spins of all the electrons as follows:

$$\rho(\vec{r}_1) = N_e \int |\psi|^2 \, d\vec{r}_2 \ldots \, d\vec{r}_{N_e} \, d\sigma_1 \ldots d\sigma_{N_e}. \qquad (8.11)$$

The quantity $\rho(\vec{r}) \, d\vec{r}$ is the number of electrons, N_e, times the probability of finding an electron in an infinitesimal volume centered at $\vec{r}$. The density $\rho(\vec{r})$ can be measured by electron diffraction and x-ray scattering experiments.

The electron density $\rho(\vec{r})$ corresponding to a Slater determinant is given by an occupation-weighted sum of squared orbitals:

$$\rho(\vec{r}) = m_1|\varphi_1(\vec{r})|^2 + m_2|\varphi_2(\vec{r})|^2 + \cdots + m_n|\varphi_n(\vec{r})|^2 \qquad (8.12)$$

where m_k is the number of electrons (1 or 2) in the kth MO. For example, the orbital wave functions for H_2^+ (equation (7.4)), He (equation (8.9)), and Li (equation (8.10)) lead to $\rho(\vec{r}) = |1\sigma_g|^2$, $\rho(\vec{r}) = 2\,|1s|^2$, and $\rho(\vec{r}) = 2\,|1s|^2 + |2s|^2$, respectively. Figure 7.2 shows ρ for the $1\sigma_g$ and $1\sigma_u$ states of H_2^+.

8.4 The Hartree–Fock model

The calculus of variations can be used to find the orbitals that result when the energy of a Slater determinant of spin orbitals is minimized using the variational method (see section 7.3). The optimal orbitals are given by the solutions of the *Hartree-Fock* (HF) *equations*, a set of equations derived by Vladimir Fock and John Slater in 1930 following earlier work by Douglas Hartree. For a closed shell atom or molecule, that is one with an even number, $2n$, of electrons paired off in n orbitals, the HF equations are

$$\hat{F}\varphi_k(\vec{r}) = \epsilon_k\varphi_k(\vec{r}) \qquad \text{for } k = 1, 2, \ldots, n, \ldots. \qquad (8.13)$$

These coupled *one*-electron equations yield the orthonormal HF orbitals φ_k as the eigenfunctions of the Fock operator $\hat{F}$. The corresponding eigenvalues ϵ_k are called *orbital energies*. Each of the n lowest-energy orbitals, called *occupied orbitals*, are combined with α and β spin functions to give the $2n$ spin orbitals used in the Slater determinant. The remaining unoccupied orbitals are called *virtual orbitals*.

The Fock operator is given by:

$$\hat{F} = -\frac{1}{2}\nabla^2 + v_{ne}(\vec{r}) + J(\vec{r}) - \hat{K} \qquad (8.14)$$

in which the first term is the kinetic energy operator for an electron, $v_{ne}(\vec{r})$ is the Coulomb attraction between an electron and each of the nuclei, and the $J(\vec{r}) - \hat{K}$ terms together account for the interelectronic repulsion. The Coulomb potential $J(\vec{r})$ is the electrostatic repulsion due to a 'smeared-out' cloud of all the electrons. Using the electron density $\rho(\vec{r})$ given by equation (8.12), it can be expressed as

$$J(\vec{r}) = \int \frac{\rho(\vec{s})}{|\vec{r} - \vec{s}|}\,d\vec{s}. \qquad (8.15)$$

The $J(\vec{r})$ potential incorrectly counts the repulsion of each electron with itself. The *exchange* operator $\hat{K}$ serves to cancel exactly the self-interaction and also includes effects arising from the Pauli antisymmetry. The internuclear repulsion V_{nn} is left out of the Fock operator because, for a fixed molecular geometry, it is a constant and can be added to the total energy at the end of the calculation; see the discussion around equation (6.4).

We need to know $\hat{F}$ to solve the HF equations (8.13) and find the HF orbitals φ_k. However, $\hat{F}$ depends on the φ_k because the Coulomb potential $J(\vec{r})$ and exchange operator $\hat{K}$ depend on the occupied φ_k. In other words, we need to know $\hat{F}$ to find the φ_k but we need to know the φ_k to find $\hat{F}$. An iterative procedure is required to overcome this mutual dependency. An initial guess of the occupied orbitals is used to generate $\rho(\vec{r})$, $J(\vec{r})$, $\hat{K}$, and thereby $\hat{F}$. Then, the HF equations are solved and the resulting orbitals used to generate an updated $\hat{F}$. This procedure is repeated until $\hat{F}$ and the electron density $\rho(\vec{r})$ are *self-consistent*—that is, they stop changing from one iteration to the next. The iterative procedure is called a *self-consistent-field*, or SCF, procedure. Symmetry arguments apply to the Fock operator in the same way as they do to the Hamiltonian (see section 4.5). Hence, permissible degeneracies of the orbitals can be determined from table 4.1.

The HF energy of a molecule is *not* the sum of the orbital energies. However, there is a nice physical interpretation of the occupied orbital energies ϵ_k. Tjalling Koopmans[2] pointed out that if one assumes the removal of a single electron from a molecule does not affect the orbitals, then a reasonable approximation to the ionization energy for removal of an electron from the orbital φ_k is given by $-\epsilon_k$. *Koopmans' approximation* is a direct connection between orbital energies and

---

[2] Koopmans asked that credit for this result be shared with his physics mentor, Hans Kramer. Koopmans began his doctoral studies in theoretical physics but later switched to mathematical economics. He shared the 1975 Nobel Memorial Prize in Economics with Leonid Kantorovich for his work on the theory of optimal resource allocation.

ionization energies measured by photoelectron spectroscopy. Relaxation of the orbitals upon ionization is included in better methods for computing ionization energies.

The method presented above for closed shell systems is sometimes referred to as the spin-restricted Hartree-Fock (RHF) method. Generalization of the RHF method to open-shell systems takes two common forms: the restricted open-shell Hartree–Fock (ROHF) and spin-unrestricted Hartree–Fock (UHF) methods. In the ROHF method, as many orbitals as possible are doubly occupied. In the UHF method, every electron is in a different orbital. For example, the ROHF wave function for Li is $|1s\overline{1s}2s|$ with one doubly occupied and one singly occupied orbital. The UHF wave function for Li is $|1s\overline{1s'}2s|$ with three singly occupied orbitals. The $1s$ and $1s'$ orbitals and their energies are very similar but not identical. Both methods have advantages and disadvantages but UHF calculations are far more common.

8.4.1 Matrix formulation

Practically exact solutions of the HF equations can be obtained for atoms and diatomic molecules by numerical grid methods. However, such methods are not always feasible for polyatomic molecules. George Hall and Clemens Roothaan transformed the integro-differential HF equations into matrix[3] equations. A *basis set* of N known real-valued functions, $\{g_1(\vec{r}), g_2(\vec{r}), \ldots, g_N(\vec{r})\}$, is used to expand each of the MOs as follows:

$$\varphi_k(\vec{r}) = c_{1k}g_1(\vec{r}) + c_{2k}g_2(\vec{r}) + \cdots + c_{Nk}g_N(\vec{r}) \tag{8.16}$$

where the real numbers c_{jk} are called MO coefficients and need to be calculated. Recall that in section 7.1 and section 7.4, the 1σ MOs of H_2^+ were expanded in basis sets of AOs on each of the hydrogen atoms. A simple derivation proves that the problem of finding the MO coefficients in equation (8.16) requires one to solve the Hall–Roothaan or *matrix Hartree–Fock equations*:

$$\mathbf{F}\mathbf{C}_k = \epsilon_k \mathbf{S}\mathbf{C}_k \qquad \text{for } k = 1, \ldots, N \tag{8.17}$$

in which $\mathbf{F}$ and $\mathbf{S}$ are $N \times N$ matrices, and $\mathbf{C}_k$ is a column vector (an $N \times 1$ matrix) that contains the expansion coefficients for the kth MO. The elements of the Fock ($\mathbf{F}$) and overlap ($\mathbf{S}$) matrices are given by

$$F_{ij} = \int g_i(\vec{r})[\hat{F}g_j(\vec{r})]\,\mathrm{d}\tau \qquad \text{and} \qquad S_{ij} = \int g_i(\vec{r})g_j(\vec{r})\,\mathrm{d}\tau. \tag{8.18}$$

The matrix equations (8.17) can be transformed into a matrix eigenvalue problem that can be solved by a computer. As on page 8-5, an iterative self-consistent-field procedure is required because the elements of the Fock matrix depend upon the occupied orbitals. Using N basis functions leads to N MOs. Hence, N must always be at least as large as, and is usually much larger than, the number of occupied MOs. As N becomes sufficiently large, the orbitals and orbital energies approach the HF limit. The choice of a suitable basis set is discussed in section 10.2.1. Some simple examples follow.

---

[3] Appendix A reviews matrix algebra.

We have already done calculations equivalent to solving the matrix HF equations for H_2^+. A matrix HF calculation using the hydrogen atom $1s_a$ and $1s_b$ AOs as basis functions would lead to 2×2 matrices in equation (8.17). Solving the HF equations would result in two MOs: the occupied MO would be $1\sigma_g$ and the virtual MO would be $1\sigma_u$, exactly as in equation (7.4). If we used the two $1s$ functions of equation (7.12) and the two $2p_z$ AOs of equation (7.14) for a total of four basis functions, then the 4×4 matrix HF equations would lead to one occupied MO as in equation (7.13) and three unoccupied MOs.

The $1s$ orbital for He in section 8.1 can be improved by a matrix HF computation using two basis functions: $s_1 = e^{-\zeta_1 r}$ and $s_2 = e^{-\zeta_2 r}$ with $\zeta_1 = 1.453$ and $\zeta_2 = 2.906$. A basis set which contains two copies of each function which differ only in their ζ values is called a *double-zeta* set. The 2×2 matrix HF equations lead to one occupied and one virtual orbital. The resulting energy $E = -2.861\,67\,E_h$ predicts an ionization energy of $I = 2262$ kJ mol^{-1} which is 111 kJ mol^{-1} smaller than experiment.

8.5 Atoms

Pioneering HF calculations on atoms were carried out in the early days of quantum chemistry by Douglas Hartree, his father William Hartree, John Slater, Bertha Swirles (later Lady Jeffreys), and others. The qualitative insights obtained by their heroic[4] calculations now appear in general chemistry books. It is convenient to work in spherical polar coordinates because the potential energy in an atom is spherically symmetric. As in the hydrogen atom, three quantum numbers (n, ℓ, and m) are needed to label the HF orbitals. Each orbital is a product of a radial function $R_{n,\ell}(r)$ and a spherical harmonic $\mathcal{Y}_{\ell,m}(\theta, \phi)$ (see section 5.2). Thus, the orbitals of any atom can be labeled $1s$, $2s$, $2p_x$, ..., just as in the hydrogen atom.

The HF orbital energies of the atoms from H to Xe are shown in figure 8.1. Note that in atoms with two or more electrons, unlike in the hydrogen atom (see figure 6.1), orbitals with the same n and different ℓ are not degenerate. For example, the $2s$ and $2p$ orbitals are degenerate in the hydrogen atom and hydrogen-like ions but they have different energies in any atom or atomic ion with two or more electrons. Observe also that the same orbital has a different energy in different atoms and atomic ions. For example, the energy of a $2p$ orbital is $-0.310\,E_h$ in B but it is $-4.256\,E_h$ in Al. This means that it is possible for the order of a pair of orbitals to change from one atom to another; for example, $3d$ orbitals are lower in energy than the $4s$ orbital in Sc, but the opposite is true in K and Ca.

It is often said that to determine the electron configuration of an atom, we move up the energy ladder filling each orbital with up to two paired electrons until all electrons have been accounted for. Degenerate orbitals are filled keeping as many unpaired electrons as possible; this is called *Hund's rule*. However, this *Aufbau* (building-up) process is too simplistic and fails even in the first transition series from Sc to Zn. As figure 8.1 shows, the $3d$ orbital energy is lower than that of $4s$ throughout this series. If strict orbital energy order was followed for Sc–Ni, then

---

[4] Keep in mind that electronic computers and calculators had not yet been invented.

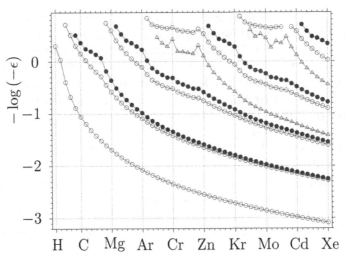

Figure 8.1. Hartree–Fock atomic orbital energies: $1s$–$5s$ ($\circ$), $2p$–$5p$ ($\bullet$), $3d$–$4d$ ($\triangle$). To identify the orbitals, note that in Xe the order is $1s$, $2s$, $2p$, $3s$, $3p$, $3d$, $4s$, $4p$, $4d$, $5s$, $5p$. Based on calculations reported by Koga T and Thakkar A J 1996 *J. Phys. B: At. Mol. Opt. Phys.* **29** 2973.

their configurations would be $[\text{Ar}]4s^0 3d^{n+2}$ with $n = 1, \ldots, 8$ where [Ar] is an abbreviation for the electron configuration of Ar. However, except for Cr and Cu, both the experimental and HF configurations for Sc–Ni are $[\text{Ar}]4s^2 3d^n$. The correct configuration is obtained only if one considers the total energies of various configurations instead of orbital energies. Recall from section 8.4 that the HF energy of a molecule is *not* the sum of the orbital energies. Approximating the total energy by the sum of the energies of the occupied orbitals can be useful, as in the Hückel model (see chapter 9), but it sometimes leads to incorrect conclusions.

As an aid to remembering the electron configurations of the atoms, one can use *Klechkowsky's rule*: fill atomic orbitals *as if it were true that* their orbital energies increase when $n + \ell$ increases, and for fixed $n + \ell$ they increase as n increases. Exceptions to this rule occur in several transition metal atoms. For example, Pd has a $[\text{Kr}]5s^0 4d^{10}$ configuration rather than $[\text{Kr}]5s^2 4d^8$ as predicted by Klechkowsky's rule.

The electron density is spherically symmetric in atoms and the radial electron density $D(r)$ is more useful as in section 6.3.3. For example, the electron configuration of Kr is $1s^2 2s^2 2p^6 3s^2 3p^6 4s^2 3d^{10} 4p^6$ and equation (8.12) leads to $\rho(r) = 2|1s|^2 + 2|2s|^2 + 6|2p|^2 + 2|3s|^2 + 6|3p|^2 + 2|4s|^2 + 10|3d|^2 + 6|4p|^2$. Figure 8.2 shows $D(r) = 4\pi r^2 \rho(r)$ for Kr. Observe that four peaks arise from the occupied orbitals with $n = 1, 2, 3,$ and 4, respectively. These peaks in $D(r)$ correspond to the K, L, M, and N shells of Kr.

The position of the outermost maximum of the orbital radial densities serves as a measure of the covalent radius. Figure 8.3 shows that these radii exhibit periodic behavior; in any given row of the periodic table, they generally decrease from left (alkali) to right (noble gas). The striking exception for Pd arises because it is the only atom in the fourth period that does not have an occupied $4s$ orbital; its electron configuration is $[\text{Kr}]4d^{10}$.

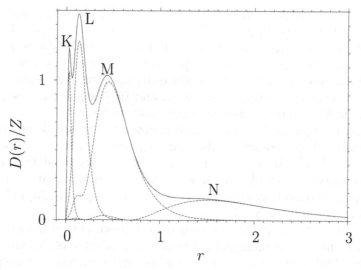

Figure 8.2. Total (——) and K-, L-, M-, and N-shell contributions (······) to the Hartree–Fock radial electron density for Kr ($Z = 36$). Calculated from the HF wave function reported by Koga T, Kanayama K, Watanabe S and Thakkar A J 1999 *Int. J. Quantum Chem.* **71** 491.

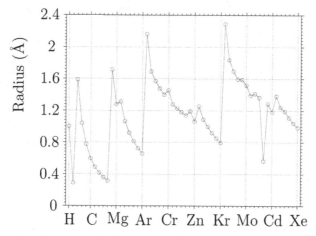

Figure 8.3. Atomic radii calculated by Waber J T and Cromer D T 1965 *J. Chem. Phys.* **42** 4116.

8.6 Diatomic molecules

The study of molecular orbitals was pioneered by Friedrich Hund, Robert Mulliken[5], and Sir John E Lennard-Jones. The insights obtained from their work now constitute some of the basic ideas of inorganic chemistry.

The HF equations for diatomic molecules can be solved numerically in confocal elliptic coordinates, (μ, ν, ϕ); see page 7-1. The molecular orbitals are written as

---

[5] Mulliken won the 1966 Nobel Prize in Chemistry for 'his fundamental work concerning chemical bonds and the electronic structure of molecules by the molecular orbital method'.

$\varphi(\mu,\nu,\phi) = f(\mu,\nu)\Phi_m(\phi)$, where $\Phi_m(\phi)$ is a particle-on-a-ring wave function (equation (5.4)) and $m = 0, \pm 1, \ldots$ for $\sigma, \pi, \ldots$ MOs, respectively. Homonuclear diatomic molecules have $D_{\infty h}$ symmetry, and their orbitals can be assigned the symmetry labels $1\sigma_g, 1\sigma_u, 2\sigma_g, 2\sigma_u, \ldots$, just as in H_2^+. Figure 8.4 shows the orbital energies of the diatomics from B_2 to F_2. The ordering of the levels is almost the same as in H_2^+ (see section 7.6) except that the $3\sigma_g$ and $1\pi_u$ energies cross twice with $1\pi_u$ being lower in B_2, C_2 and F_2, and $3\sigma_g$ being lower in N_2 and O_2.

The content of figure 8.4 can be summarized by writing down the electron configurations. For example, the configuration of C_2 is written as $1\sigma_g^2 1\sigma_u^2 2\sigma_g^2 2\sigma_u^2 1\pi_{ux}^2 1\pi_{uy}^2$. It is common to use $1\pi_u^4$ as shorthand for $1\pi_{ux}^2 1\pi_{uy}^2$. In inorganic chemistry textbooks, the numbering is sometimes changed to omit the MOs that contain the core electrons; as a consequence, $1\sigma_g$ in H_2, $2\sigma_g$ in Li_2, and $3\sigma_g$ in N_2 may all be confusingly called $1\sigma_g$.

The B_2 and O_2 molecules have two unpaired electrons in their highest occupied π MOs and are therefore paramagnetic. The correct prediction of the paramagnetism of O_2 by the MO theory was one of its earliest major successes. A nominal *bond order* is given by the number of bonding electron pairs minus the number of

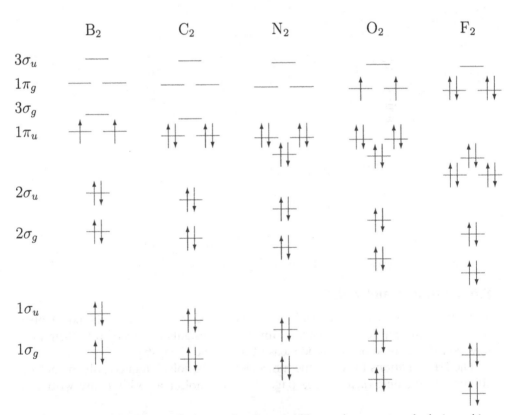

Figure 8.4. MO energy level diagram for homonuclear diatomics. The energies are *not* to scale; the $1\sigma_g$ and $1\sigma_u$ energies are far lower than the energies of the valence MOs.

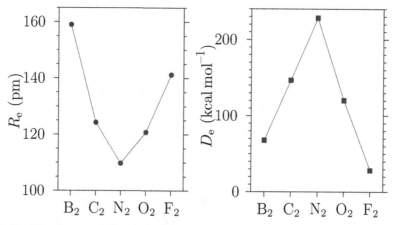

Figure 8.5. Bond lengths R_e and dissociation energies D_e. The nominal bond order is 1, 2, 3, 2, and 1 for B_2, C_2, N_2, O_2, and F_2, respectively.

antibonding electron pairs. Bond orders of 1, 2, and 3 correspond to single, double, and triple bonds, respectively. Figure 8.5 shows that, for the second-period A_2 molecules, the greater the bond order, the shorter the bond length and the greater the dissociation energy. Figure 8.4 suggests that Be_2 and Ne_2 have a bond order of zero and are not stable molecules. In reality, these molecules are very weakly bound by van der Waals forces.

8.7 The Kohn–Sham model

The HF model is useful but also has severe limitations. In section 8.4 we saw that there is a residual error of $111 \, \text{kcal mol}^{-1}$ in the ionization energy of the helium atom. Further refinement of the He orbital by using a basis set of three or more exponential (Slater-type) functions does not help. A 5×5 matrix HF calculation of this type leads to an energy $E = -2.861\,68\,E_h$, which is not much better than the energy obtained with two functions. Unfortunately, this is the best energy that can be obtained with a HF wave function. Moreover, the breaking of a bond is very poorly described by the HF model if *both* the resulting fragments are open-shell species. For example, figure 8.6 shows that the HF potential energy curve for H_2 diverges from the exact one as the bond is broken.

The HF model wave function is unable to describe the instantaneous, short-range, *dynamical correlation* of the motion of the electrons that keeps them apart. It helps to think of electron correlation as creating a *Coulomb hole* or bubble around each electron in which the probability of finding another electron with opposite spin is greatly reduced. We now turn to the Kohn–Sham (KS) model because it improves upon the HF model and creates a Coulomb hole around each electron. The KS model stems from the 1951 work of John Slater who was trying to simplify the HF model. He replaced the exchange operator $\hat{K}$ in the HF equations (8.13) by an effective exchange potential of the form $v_x(\vec{r}) = c\,\rho(\vec{r})^{1/3}$, where c is a constant. To understand the idea

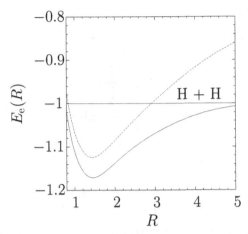

Figure 8.6. Potential energy curves, in atomic units, for the ground electronic state of H_2. -----: Hartree–Fock; ——: nearly exact.

behind v_x, observe that the electron density ρ has dimensions of a reciprocal volume and hence $\mathcal{R} = \rho^{-1/3}$ is a length. Now rewrite the exchange potential as $v_x = c/\mathcal{R}$ and note that v_x has the same general form as the Coulomb potential. Now visualize $\mathcal{R}$ as the radius of an exclusion sphere around an electron at the position $\vec{r}$. The radius $\mathcal{R}$ shrinks in regions of high density and expands in regions of low density. This is physically quite reasonable; compare the size of your personal space when you are in a crowded theater with its size when you are on a deserted beach.

Slater's approach unexpectedly improves upon HF results in some cases, suggesting that v_x partially models electron correlation. So why not replace the exchange operator $\hat{K}$ in the HF equations by an *exchange-correlation potential* $v_{xc}(\vec{r})$ that mimics both the role of $\hat{K}$ and the effects of electron correlation? This line of inquiry is called Kohn–Sham *density functional theory* (KS-DFT) because of an important formal development by Walter Kohn[6] and Lu Jeu Sham following earlier work by Pierre Hohenberg and Kohn. Their work guarantees the existence of an exact v_{xc}. Unfortunately, the exact v_{xc} is unknown and approximate forms must be used. KS–DFT is being developed very rapidly at this time and there are many competing v_{xc} models, some of which will be discussed in section 10.2.2.

Matrix KS equations analogous to the matrix HF equations are obtained by modifying the Fock operator in equation (8.18) appropriately. The KS MOs resemble, but differ from, HF MOs. They have the same symmetry properties as their HF counterparts. The KS energy of the highest occupied MO equals the first

---

[6] Walter Kohn shared the 1998 Nobel Prize in Chemistry for his development of density functional theory. Many people loosely refer to KS–DFT as simply DFT, even though DFT should be used to denote a pure density-only theory.

ionization energy of the molecule. Using KS orbitals in equation (8.12) gives the electron density. However, the Slater determinant of KS orbitals is *not* an accurate wave function for the system of interest. Hence, it is not known how to use KS–DFT correctly for molecular properties that depend on quantities other than the energy and electron density.

Problems (see appendix B for hints and solutions)

8.1 Using atomic units, write down the electronic Hamiltonian for the Li^+ cation. Explain the physical meaning of each term.

8.2 Why is it wrong to use $[1s(1)\overline{1s}(2) + \overline{1s}(1)1s(2)]/\sqrt{2}$ as an approximate wave function for the ground state of the helium atom?

8.3 A Slater determinant satisfies the Pauli principle. So what is wrong with using $|1s\overline{1s}1s|$ for the ground state of the Be^+ cation?

8.4 Consider an excited (triplet) state of the helium atom which has the configuration $1s2s$ and both electrons have β spin. Set up an orbital-model wave function that satisfies the Pauli postulate for this excited state. Write your wave function in the form of (a) a 2×2 determinant, and (b) a spatial function multiplied by a spin function.

8.5 In the compact notation of page 8-3, write down a Slater determinant which could be used to represent the ground-state wave function of the boron atom in (a) the ROHF method, and (b) the UHF method.

8.6 The energies of the occupied molecular orbitals, $1a_1$, $2a_1$, and the triply degenerate $1t_2$, of CH_4 are −11.20, −0.93 and −0.54 E_h, respectively. Estimate the ionization energy of CH_4.

8.7 Consider the ground state of a system containing two electrons and a nucleus of atomic number Z. If we use the Slater determinant of equation (8.9) with the hydrogen-like $1s$ function of equation (8.3) as the trial wave function in a variational calculation, we find $E = \zeta^2 - 2Z\zeta + 5\zeta/8$.
(a) Find the ζ that should be used *without* using a specific value for Z. Justify your answer. For He ($Z = 2$), your answer should agree with the corresponding result in section 8.1.
(b) What is the value of the ground-state energy predicted if the above ζ is used? Do *not* use a specific value for Z at this stage.
(c) Compare the ground-state energy of the H^- anion ($Z = 1$) predicted by this calculation with that of the hydrogen atom. What does this lead you to expect about the stability of H^-?

8.8 How many peaks should one see in the photoelectron spectrum of the Ar atom? Explain your answer.

8.9 Use figure 8.4 to write the ground-state electron configurations of O_2 and O_2^+. Which should have the longer bond length and why?

8.10 Write the ground-state electron configuration of Li_2. Draw the MO energy-level diagram and sketch the occupied MOs. What is the nominal bond order? Would Li_2^+ be more or less strongly bound than Li_2?

8.11 Write the ground-state electron configuration of Be_2. Draw the MO energy-level diagram and sketch the occupied MOs. What is the nominal bond order? What is the lowest unoccupied molecular orbital (LUMO)?

IOP Concise Physics

Quantum Chemistry
A concise introduction for students of physics, chemistry, biochemistry and materials science
Ajit J Thakkar

Chapter 9

Qualitative MO theory

9.1 The Hückel model

In 1930, Erich Hückel proposed a simple model for the π MOs and energies[1] of planar conjugated hydrocarbons; the insights his model provides have permeated organic chemistry. The Hückel model remains important although far more sophisticated calculations are possible now.

The Hückel model is 'obtained' from the HF model as follows.

1. All σ electrons are neglected except insofar as they affect parameters of the method. This means that all hydrogen atoms are ignored.

2. The basis set for a planar hydrocarbon with N carbon atoms is chosen to be one $2p_\pi$ AO on each carbon atom. If the hydrocarbon is in the xy-plane, these are the $2p_z$ AOs. Thus the kth MO is given by

$$\varphi_k = c_{1k}p_1 + c_{2k}p_2 + \cdots + c_{Nk}p_N \tag{9.1}$$

where p_i is the AO on carbon i, and the c_{ik} are expansion coefficients.

3. The elements of the **F** and **S** matrices appearing in the matrix HF equations (8.17) are assigned values instead of being calculated. All overlap is neglected by setting $S_{ij} = 1$ when $i = j$, and $S_{ij} = 0$ otherwise[2]. Thus, the Hückel overlap matrix is an $N \times N$ identity matrix **I**; that is **S** = **I**. This implies that the normalization condition for the Hückel MO of equation (9.1) is simply

$$c_{1k}^2 + c_{2k}^2 + \cdots + c_{Nk}^2 = 1. \tag{9.2}$$

---

[1] The MOs of planar molecules are called σ or π, respectively, depending on whether they remain unchanged or change sign upon reflection in the molecular plane. Unfortunately, this usage is entrenched in chemistry although it is inconsistent with symmetry conventions. Note that degeneracy is *not* implied by the π label in this context.

[2] In fact, if the carbon atoms i and j are separated by a typical CC bond distance, then the overlap between their $2p_\pi$ orbitals is $S_{ij} \approx 0.25$. However, inclusion of overlap does not change the results by huge amounts.

doi:10.1088/978-1-627-05416-4ch9

4. A diagonal Fock matrix element F_{ii} is roughly the energy of an electron in the $2p_\pi$ AO of carbon i. They are all made equal because the carbon atoms in conjugated hydrocarbons are nearly equivalent: $F_{ii} \approx \alpha$. Thus, the diagonal part of the Hückel F is simply αI.

5. If carbon atoms i and j are not bonded to each other, their interaction is small, and the matrix elements $F_{ij} = F_{ji} \approx 0$. If the carbon atoms i and j are bonded to each other, then Hückel sets the matrix elements $F_{ij} = F_{ji}$ to an interaction energy, β. Moreover, this negative value β is used for all bonded pairs because all carbon atoms in conjugated hydrocarbons are nearly equivalent and all CC bond lengths are nearly equal. Thus, the off-diagonal part of the Hückel F matrix is βA in which the elements of A are given by $A_{ij} = 1$ if atoms i and j are bonded and by $A_{ij} = 0$ otherwise. A is called the *adjacency matrix* because the only information it contains is the connectivity of the molecule—which carbons are bonded to which other carbons. Putting the diagonal and off-diagonal parts together, the Fock matrix becomes $F = \alpha I + \beta A$. The Hückel F reflects the permutational symmetry of A but does *not* contain any geometrical information. For example, in cyclobutadiene with atoms numbered sequentially,

$$F = \begin{bmatrix} \alpha & \beta & 0 & \beta \\ \beta & \alpha & \beta & 0 \\ 0 & \beta & \alpha & \beta \\ \beta & 0 & \beta & \alpha \end{bmatrix} = \alpha \begin{bmatrix} 1 & 0 & 0 & 0 \\ 0 & 1 & 0 & 0 \\ 0 & 0 & 1 & 0 \\ 0 & 0 & 0 & 1 \end{bmatrix} + \beta \begin{bmatrix} 0 & 1 & 0 & 1 \\ 1 & 0 & 1 & 0 \\ 0 & 1 & 0 & 1 \\ 1 & 0 & 1 & 0 \end{bmatrix}.$$

6. The π-electron energy E_π is taken as the sum of the orbital energies.

Inserting the Hückel S and F into the matrix equations (8.17) leads to

$$(\alpha I + \beta A)C_k = \epsilon_k I C_k \qquad \text{for } k = 1, \dots, N. \qquad (9.3)$$

Subtracting $\alpha I C_k$ from both sides of equation (9.3), using $I C_k = C_k$, collecting terms, and dividing both sides by β, we find

$$A C_k = x_k C_k \qquad \text{for } k = 1, \dots, N \qquad (9.4)$$

in which $x_k = (\epsilon_k - \alpha)/\beta$. F and A have the same eigenvectors and the Hückel orbital energies are obtained from the eigenvalues of A using

$$\epsilon_k = \alpha + x_k \beta \qquad \text{for } k = 1, \dots, N. \qquad (9.5)$$

Equation (9.4) provides a route to the Hückel orbital energies and MOs *without* the need to assign numerical values to α and β.

9.2 Cumulenes

A *cumulene* is a conjugated non-branched chain of N carbons, $H(CH)_N H$.

Ethene, the allyl radical, and 1,3-butadiene are cumulenes with $N = 2$, 3, and 4, respectively. The Hückel energies and MOs are shown in figure 9.1 for ethene. As in chapter 7, the positive lobes of the MOs are shaded. φ_1 is a bonding MO because it leads to a build-up of charge between the nuclei, whereas φ_2 has a nodal plane perpendicular to the C=C bond and is antibonding. Hence, the orbital energy $\alpha + \beta$ is the lower one confirming that $\beta < 0$. The total π-electron energy is $E_\pi = 2\epsilon_1 = 2(\alpha + \beta)$.

Next consider the orbital energies shown in figure 9.2 for butadiene. The four π-electrons occupy the two lowest energy levels. Thus, $E_\pi = 2\epsilon_1 + 2\epsilon_2 = 4\alpha + 4.48\beta$. If the two π-bonds were non-interacting, then we would expect a π-electron energy twice that of ethene—that is, $4(\alpha + \beta)$. The difference between the Hückel E_π and its hypothetical non-interacting counterpart is called the *delocalization energy*, ΔE_{dl}. For butadiene, $\Delta E_{dl} = E_\pi - 4(\alpha + \beta) = 0.48\beta$. The delocalization energy is due to *conjugation*.

The sketches of the MOs in figure 9.2 follow from the MO coefficients noting that the carbon atoms are numbered from left to right. The shaded lobes of the atomic p orbitals are positive and the unshaded ones are negative. The size of each atomic orbital is proportional to its coefficient in that MO. The lowest-energy MO (φ_1) is bonding between each pair of adjacent carbons. The next MO, φ_2, has two bonding interactions (between carbons 1–2 and 3–4) and one antibonding interaction

$$\epsilon_2 = \alpha - \beta \qquad \underline{\quad\quad} \qquad \varphi_2 = 0.71\,(p_1 - p_2)$$

$$\epsilon_1 = \alpha + \beta \qquad \underline{\uparrow\downarrow} \qquad \varphi_1 = 0.71\,(p_1 + p_2)$$

Figure 9.1. Hückel energy levels and MOs for ethene.

$$\epsilon_4 = \alpha - 1.62\beta \qquad \underline{\quad\quad} \qquad \varphi_4 = 0.37p_1 - 0.60p_2 + 0.60p_3 - 0.37p_4$$

$$\epsilon_3 = \alpha - 0.62\beta \qquad \underline{\quad\quad} \qquad \varphi_3 = 0.60p_1 - 0.37p_2 - 0.37p_3 + 0.60p_4$$

$$\epsilon_2 = \alpha + 0.62\beta \qquad \underline{\uparrow\downarrow} \qquad \varphi_2 = 0.60p_1 + 0.37p_2 - 0.37p_3 - 0.60p_4$$

$$\epsilon_1 = \alpha + 1.62\beta \qquad \underline{\uparrow\downarrow} \qquad \varphi_1 = 0.37p_1 + 0.60p_2 + 0.60p_3 + 0.37p_4$$

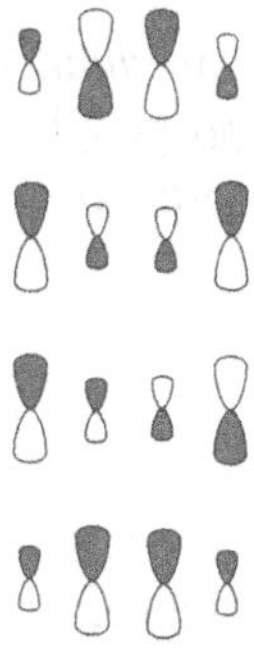

Figure 9.2. Hückel energies and MOs for 1,3-butadiene.

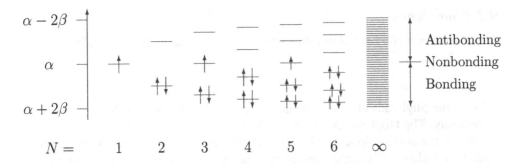

Figure 9.3. Hückel energy levels for cumulenes.

between carbons 2–3, which has the same net effect as one bonding interaction. The third MO has one bonding and two antibonding interactions, and φ_4 has three antibonding interactions. The energy increases as the number of nodal planes perpendicular to the molecular plane increases from 0 in the lowest energy MO φ_1 to 3 in the highest-energy MO φ_4. A smooth curve that connects the tops of the shaded lobes in a Hückel MO of ethene or butadiene (or any other cumulene) looks like a wave function of a particle in a wire. This can often be used to sketch a Hückel MO for a cumulene without calculation.

The adjacency matrix for any cumulene has ones on the first super- and sub-diagonals and zeroes elsewhere. This simplicity allows general solutions to be found. As seen in figure 9.3, the Hückel energy levels for the cumulenes are non-degenerate, and lie between $\alpha + 2\beta$ and $\alpha - 2\beta$. Every energy level at $\alpha + x\beta$ has a counterpart at $\alpha - x\beta$. If N is odd, there is a non-bonding level at $\epsilon = \alpha$. The interactions among the N atomic orbitals 'spread' an energy level into N levels.

The Hückel model can be applied to chains of N sodium atoms taking the N basis functions to be the outermost $3s$ orbitals on each of the Na atoms. Hence, a very long cumulene may also be regarded as a model of a one-dimensional strip of metallic sodium or any other alkali metal. As $N \to \infty$, the energy levels get packed into a continuous band of width $4|\beta|$. Moreover, the bandwidth increases with the strength of the interaction $|\beta|$. These ideas are of great importance in the study of solids.

9.3 Annulenes

An *annulene* is a monocyclic conjugated polyene, $C_N H_N$, with $N \geqslant 3$. The cyclo-propenyl radical, cyclobutadiene, and benzene are annulenes with $N = 3$, 4, and 6, respectively. Joining the two ends of a cumulene gives an annulene.

Hence, the adjacency matrix for an N-annulene is obtained from A for an N-cumulene by setting $A_{1N} = A_{N1} = 1$. The simplicity of A allows a general solution

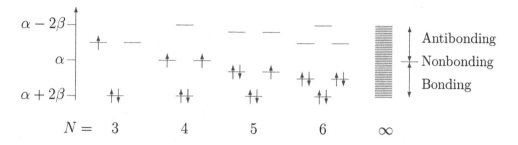

$N = \quad 3 \qquad 4 \qquad 5 \qquad 6 \qquad \infty$

Figure 9.4. Hückel energy levels for annulenes.

to be found. Note from figure 9.4 that all the annulenes have a non-degenerate level at $\alpha + 2\beta$ followed by doubly degenerate levels; when N is even, there is also a non-degenerate level at $\alpha - 2\beta$. When N is even, every energy level at $\alpha + x\beta$ has a counterpart at $\alpha - x\beta$. This pairing of energy levels does not occur in annulenes with odd N. Observe that as $N \to \infty$, the doubly degenerate energy levels get packed into a continuous band of width $4|\beta|$.

There is a striking similarity between the Hückel orbital energies for annulenes and the energy levels of a particle on a ring (see figure 5.2 on page 5-3). To have a stable filled-subshell π-electron configuration, the number of π electrons must be $N_\pi = 2 + 4m$ for $m = 0, 1, \ldots$. This gives *Hückel's $4m + 2$ rule*: monocyclic conjugated systems are most stable if they contain $4m + 2$ π electrons. Benzene and the cyclopentadienyl anion satisfy the $4m + 2$ rule with $m = 1$ because they have six π electrons. Such molecules are called *aromatic*. The Hückel model predicts that annulenes with $4m$ π electrons are highly unstable diradicals. Monocyclic conjugated hydrocarbons with $4m$ π electrons are called anti-aromatic.

The Hückel orbital energies and MOs of benzene are shown in figure 9.5 where the carbons are numbered sequentially around the ring. The lowest energy MO φ_1 is bonding between each pair of adjacent carbons. The φ_2 MO has four bonding interactions (between carbons 1–2, 1–6, 3–4, and 4–5) and two anti-bonding interactions (between carbons 2–3 and 5–6) which has the same net effect as two bonding interactions. Its degenerate partner φ_3 has two bonding interactions (between carbons 2–3 and 5–6) and no antibonding interactions. The degenerate MOs, φ_4 and φ_5, have one nodal plane perpendicular to the molecular plane. The MO energy increases with increasing number of nodal planes perpendicular to the molecular plane; there are 0 nodal planes of this type in φ_1 and 3 in the $k = 6$ MO.

The Hückel π-electron energy for benzene is $E_\pi = 2(\alpha + 2\beta) + 4(\alpha + \beta) = 6\alpha + 8\beta$. If the three π bonds were non-interacting, then we would expect a π-electron energy three times that of ethene—that is, $6(\alpha + \beta)$. The delocalization energy is then $\Delta E_{dl} = 6\alpha + 8\beta - 6(\alpha + \beta) = 2\beta$. This large delocalization energy gives benzene an extra stability. It is precisely this stability that is predicted by Hückel's $4m + 2$ rule.

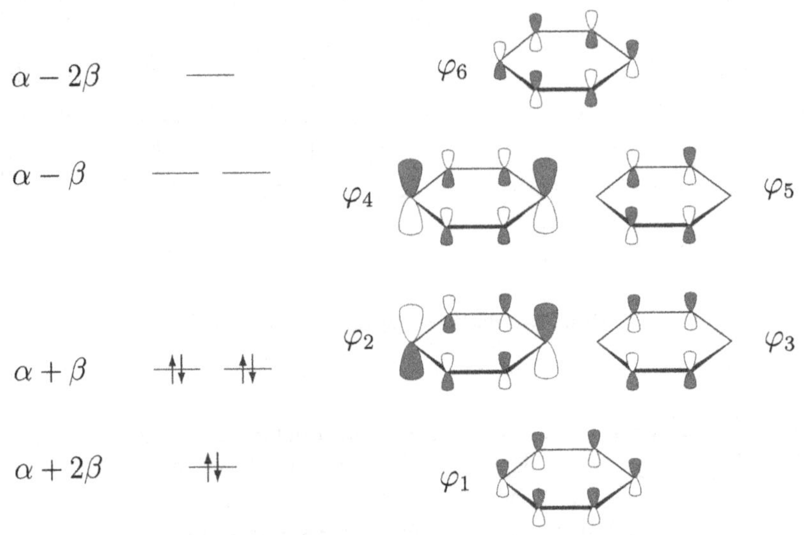

$$\varphi_6 = (p_1 - p_2 + p_3 - p_4 + p_5 - p_6)/\sqrt{6}$$
$$\varphi_5 = (p_2 - p_3 + p_5 - p_6)/2$$
$$\varphi_4 = (2p_1 - p_2 - p_3 + 2p_4 - p_5 - p_6)/\sqrt{12}$$
$$\varphi_3 = (p_2 + p_3 - p_5 - p_6)/2$$
$$\varphi_2 = (2p_1 + p_2 - p_3 - 2p_4 - p_5 + p_6)/\sqrt{12}$$
$$\varphi_1 = (p_1 + p_2 + p_3 + p_4 + p_5 + p_6)/\sqrt{6}$$

Figure 9.5. Hückel energy levels and MOs for benzene.

9.4 Other planar conjugated hydrocarbons

We now turn to other hydrocarbons. The Hückel energy levels for planar conjugated hydrocarbons with one or more tertiary carbon atoms bonded to three other carbons do not all lie between $\alpha + 2\beta$ and $\alpha - 2\beta$; for example, the lowest orbital energy for styrene (phenylethene) is at $\alpha + 2.14\beta$. However, it is true that the Hückel energy levels for *any* planar conjugated hydrocarbon always lie between $\alpha + 3\beta$ and $\alpha - 3\beta$.

Conjugated hydrocarbons which contain a ring with an odd number of atoms are said to be non-alternant. All other conjugated hydrocarbons are called *alternant hydrocarbons*. For any alternant hydrocarbon, it is always possible to place labels on some of the carbons in such a manner that no two neighbors are both labeled or both unlabeled. Such a labeling is impossible for a non-alternant hydrocarbon.

The *pairing theorem* states that for each Hückel energy level at $\alpha + x\beta$, an alternant hydrocarbon has another one at $\alpha - x\beta$. We have seen this for the cumulenes in section 9.2 and for the alternant annulenes (those with an even number N of carbon atoms) in section 9.3. A simple consequence of the pairing theorem is

$$\alpha - 1.48\beta \quad \underline{} \quad \varphi_4 = 0.30p_1 + 0.30p_2 - 0.75p_3 + 0.51p_4$$
$$\alpha - 1.00\beta \quad \underline{} \quad \varphi_3 = 0.71p_1 - 0.71p_2$$

$$\alpha + 0.31\beta \quad \underline{\uparrow\downarrow} \quad \varphi_2 = 0.37p_1 + 0.37p_2 - 0.25p_3 - 0.82p_4$$

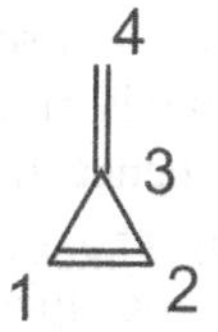

$$\alpha + 2.17\beta \quad \underline{\uparrow\downarrow} \quad \varphi_1 = 0.52p_1 + 0.52p_2 + 0.61p_3 + 0.28p_4$$

Figure 9.6. Hückel energies and MOs for methylenecyclopropene.

that the sum of all the Hückel energies, occupied and unoccupied, is $N\alpha$ for alternant hydrocarbons with N carbon atoms. This result is also true for non-alternant hydrocarbons even though the pairing theorem is inapplicable; see problem 9.10.

Several computer programs are freely available for performing numerical calculations to obtain detailed solutions of the Hückel model for conjugated hydrocarbons other than cumulenes and annulenes. As an example, figure 9.6 shows the results of a computer calculation on methylenecyclopropene. The pairing theorem does not apply to this molecule because it is a non-alternant hydrocarbon. Hence the energy levels are not placed symmetrically with respect to α. Nevertheless, the sum of all four orbital energies is precisely 4α. Carbon atom 3 is bonded to three other carbons and hence the lowest energy level is lower than $\alpha + 2\beta$, unlike in any cumulene or annulene. The two lowest energy levels are doubly occupied and the total π energy is $E_\pi = 4\alpha + 4.96\beta$ corresponding to a delocalization energy of $\Delta E_{dl} = E_\pi - N_\pi(\alpha + \beta) = 0.96\beta$. The delocalization energy is due to conjugation. φ_1 is bonding over the entire molecule. φ_2 has bonding character between carbons 1–2 and 3–4 but is antibonding between carbons 1–3 and 2–3, making it almost non-bonding overall as reflected by the closeness of its energy to α.

9.5 Charges, bond orders, and reactivity

Each term on the left-hand side of the normalization condition, equation (9.2), for Hückel MO φ_k pertains to a different carbon atom; one may interpret $|c_{jk}|^2$ as the fraction of an electron in φ_k that resides on atom j. Adding up a contribution from each MO, weighted by the number of electrons in it, gives the π-*electron population* (or π *charge*) q_j of atom j. Thus, we write

$$q_j = m_1|c_{j1}|^2 + m_2|c_{j2}|^2 + \cdots + m_N|c_{jN}|^2 \tag{9.6}$$

in which m_k is the number of electrons in the kth MO and c_{jk} is the coefficient of p_j in the kth MO. For example, the population on atom 4 in methylenecyclopropene, figure 9.6, is $q_4 = 2 \times 0.28^2 + 2 \times (-0.82)^2 = 1.49$ electrons. Similarly, we find that $q_1 = q_2 = 0.82$ and $q_3 = 0.88$ electrons. Obviously, unoccupied MOs do not contribute to q_j because they have $m_k = 0$. The sum of the populations on all the atoms equals the total number of π electrons in the molecule (four in this case).

The π-electron populations are indicators of reactivity. An electrophile is most likely to attack the carbon with the largest π charge whereas a nucleophile is most likely to attack the carbon with the smallest π charge. The π-electron populations calculated above for methylenecyclopropene show a build-up of electron density at the methylene carbon, making it a prime target for electrophilic attack.

Unfortunately, π charges are useless as reactivity indices for alternant hydrocarbons because all the carbons have the same π population. An electrophile is most likely to seek electrons from the highest occupied MO (HOMO), and hence the relevant measure of reactivity is the π charge in the HOMO given by $c_{j,\text{HOMO}}^2$. Similarly, a nucleophile is most likely to donate electrons to the lowest unoccupied MO (LUMO) and the reactivity of carbon j to nucleophilic attack can be measured by $c_{j,\text{LUMO}}^2$. These reactivity indices based on the frontier orbitals, the HOMO and LUMO, are also used as additional reactivity indices for non-alternant hydrocarbons.

Charles Coulson generalized equation (9.6) by defining the π-*bond order* of the bond between a pair of bonded carbon atoms i and j as follows:

$$P_{ij} = m_1 c_{i1} c_{j1} + m_2 c_{i2} c_{j2} + \cdots + m_N c_{iN} c_{jN}. \tag{9.7}$$

This definition leads to a π-bond order of 1 in ethene as expected. The Lewis structure for butadiene has perfectly localized, alternating double and single bonds; in other words, it shows π-bond orders of $P_{12} = 1$, $P_{23} = 0$ and $P_{34} = 1$. By contrast, substituting MO coefficients from figure 9.2 into equation (9.7), the Hückel π-bond orders are found to be $P_{12} = 2 \times 0.37 \times 0.60 + 2 \times 0.60 \times 0.37 = 0.89$, $P_{23} = 2 \times 0.60 \times 0.60 + 2 \times 0.37 \times (-0.37) = 0.45$, and $P_{34} = 0.89$. Comparison with the Lewis bond orders shows that the bond between the central carbons has acquired some π character at the expense of the terminal bonds; in other words, the π electrons are delocalized due to conjugation. General chemistry textbook descriptions of the 'resonance' between two equivalent Kekulé structures suggest that the CC π-bond orders in benzene should be close to $1/2$. The Hückel model predicts the CC π-bond orders in benzene to be $2/3$ indicating that π-electron *delocalization* strengthens all the CC bonds. The π-bond order between carbon atoms 1 and 3 in methylenecyclopropene is $P_{13} = 0.45$ showing π-electron delocalization in the three-membered ring. The total bond order is $1 + P_{ij}$ since there is also a σ bond between every pair of bonded carbons.

Molecular geometry plays no role in the Hückel model because it uses only connectivity information. For example, there is no difference between cis- and trans-butadiene in the Hückel model because both isomers have the same connectivity. However, Coulson found a relationship between Hückel model π-bond orders P_{jk} and experimental bond lengths R_{jk}:

$$R_{jk} \approx R_s - \frac{P_{jk}(R_s - R_d)}{P_{jk} + 0.765(1 - P_{jk})} \tag{9.8}$$

in which $R_s = 154$ pm and $R_d = 134$ pm are typical CC single and double bond lengths, respectively. Equation (9.8) is set up to predict correctly that $R_{jk} = R_s$ when $P_{jk} = 0$ and $R_{jk} = R_d$ when $P_{jk} = 1$. This simple empirical formula gives useful predictions of CC bond lengths in conjugated hydrocarbons. For example, it correctly predicts CC bond lengths of 140 pm in benzene. Formula (9.8) gives a C2–C3 bond length of 144 pm in butadiene to be compared with the experimental value of 146 pm.

Reaction with a free radical should proceed with the radical forming a bond to a carbon. This will be easiest with carbons which have the greatest bonding capacity left over after accounting for the bonds that are initially present in the molecule. Thus, the *free valence* is the difference between the maximum π bonding a carbon atom can have and the amount of π bonding it actually has in the molecule prior to reaction. The amount of π bonding (B_k) at atom k is measured by the sum of the orders of all the π bonds involving the atom. Coulson defined the free valence index of carbon k as

$$f_k = \sqrt{3} - B_k \qquad (9.9)$$

in which $\sqrt{3}$ is the largest possible Hückel value[3] of B_k for any trigonally bonded carbon. For example, using the π-bond orders calculated on page 9-8, the free valence of carbon atom 1 in butadiene is $f_1 = \sqrt{3} - P_{12} = 1.73 - 0.89 = 0.84$ whereas the carbon atom 2 has $f_2 = \sqrt{3} - P_{12} - P_{23} = 1.73 - 0.89 - 0.45 = 0.39$. Check that $f_3 = f_2$ and $f_4 = f_1$. The free valence values correctly predict that the primary carbons are more vulnerable to free radical attack than the secondary carbons.

By and large, the various reactivity indices provide very simple and useful measures of reactivity. However, none of these reactivity indices is perfect. Sometimes the different indices predict different things. One must expect a model as simple as the Hückel model to have shortcomings.

9.6 The Hückel model is not quantitative

The Hückel method provides very useful qualitative results for conjugated hydrocarbons. The $4m + 2$ rule is an outstanding example of its success. Unfortunately, attempts to use the Hückel model quantitatively by assigning numerical values to α and β revealed that different values are required for different properties to agree with experiment. Hence, quantitative use of the Hückel model has been abandoned.

An extended Hückel method that includes all valence electrons plays an analogous role for molecules of all types, organic and inorganic. The method provides qualitative results such as the nodal structure and symmetries of molecular orbitals. The Woodward–Hoffmann[4] rules used in organic chemistry were based on such calculations. It is not, however, a quantitative method like those discussed in chapter 10.

---

[3] The maximal value of $B_k = \sqrt{3}$, and hence $f_k = 0$, is found for the central carbon atom in the diradical $C(CH_2)_3$.

[4] Robert Woodward won the 1965 Nobel Prize in Chemistry for his work on organic synthesis, and Kenichi Fukui and Roald Hoffmann shared the 1981 Nobel Prize in Chemistry for their work on quantum theories of chemical reactions.

Problems (see appendix B for hints and solutions)

9.1 Write down the adjacency matrix for (a) butadiene (C_4H_6) with the atoms numbered sequentially from left to right, and (b) benzene (C_6H_6) with the atoms numbered sequentially around the ring.

9.2 Prove that the orthogonality condition $\int \varphi_i \varphi_j \, d\tau = 0$ for $i \neq j$ can be expressed in terms of the Hückel MO coefficients as

$$c_{1i}c_{1j} + c_{2i}c_{2j} + \cdots + c_{Ni}c_{Nj} = 0 \qquad \text{for } i \neq j.$$

9.3 Show that the HOMO and LUMO in figure 9.2 are orthonormal.

9.4 Use results from section 9.2 and section 9.3 to draw a π-orbital energy level diagram for (a) cyclo-octatetraene and (b) octatetraene.

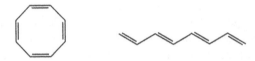

Which of these molecules would you expect to be more stable? Why?

9.5 Which of $C_5H_5^+$, C_5H_5 and $C_5H_5^-$ is expected to be the most stable? How does this help to understand the structure of ferrocene, $Fe(C_5H_5)_2$?

9.6 Given $\alpha = -6.15$ eV and $\beta = -3.32$ eV, predict the ionization energy for benzene and compare it with the experimental value of 9.4 eV.

9.7 A student performed a Hückel calculation on fulvene.
(a) She found that C3, C4, and C5 had π-electron populations of 1.073, 1.092, and 1.047, respectively. What were the π-electron populations on C1, C2, and C6?

(b) Use these π-electron populations to predict which carbon atom a nucleophile is most likely to attack. Explain your choice.
(c) The student found π-bond orders of 0.76, 0.45, 0.78, and 0.52 for the C6–C5, C5–C4, C4–C3, and C3–C2 bonds, respectively. Calculate the free valence index for C2.
(d) Use π-bond orders and free valence indexes to predict which carbon atom a free radical would be most likely to attack. Explain your choice.

9.8 A lone-pair electron on the O atom in $H_2C{=}CH{-}CH{=}O$ is promoted to the lowest unoccupied molecular orbital (LUMO). Assuming that the π-MOs of prop-2-enone are the same as in butadiene, explain with MO diagrams but *without calculations* which CC bond you would expect to become longer and which shorter as a result of this transition, and why.

9.9 Calculate the π-bond orders of the C1–C2 and C2–C3 bonds for butadiene using equation (9.7) and the Hückel MOs given in figure 9.2. Check against section 9.5. Suppose an electron moves from the π HOMO to the π LUMO in butadiene. Find the π-bond orders for this excited state.

9.10 Calculate the trace (sum of diagonal elements) of the Hückel **F** matrix. What does this tell us about the energy levels?

9.11 Alice found a pair of degenerate Hückel MOs for cyclobutadiene:

$$\varphi_a = (p_1 - p_3)/\sqrt{2} \qquad \text{and} \qquad \varphi_b = (p_2 - p_4)/\sqrt{2}.$$

However, Jim thought that the same pair of degenerate MOs were given by:

$$\varphi_c = (p_1 - p_2 - p_3 + p_4)/2 \qquad \text{and} \qquad \varphi_d = (p_1 + p_2 - p_3 - p_4)/2.$$

They were both right. Explain how that can be true.

Quantum Chemistry
A concise introduction for students of physics, chemistry, biochemistry and materials science
Ajit J Thakkar

Chapter 10

Computational chemistry

10.1 Computations are now routine

Computations complement experimental work. They are used to predict quantities that are too hard, expensive, time-consuming, or hazardous to measure. For example, the structure of a transition state is easier to calculate than measure. Calculations can guide the design of experimental work. For example, pharmaceutical companies developing new drugs routinely use computational methods to screen compounds for likely activity before synthesizing them. Computational methods can predict equilibrium geometries, transition-state structures, dipole moments, barriers to internal rotation, relative energies of isomers, enthalpies of reaction and formation, and many other properties. It is a rare chemical problem where computational chemistry is of no help at all.

The HF and KS methods described in chapter 8 have been implemented in many computer programs, such as GAMESS, ORCA and GAUSSIAN. The programs can be run on cheap commodity computers. Virtually anyone can easily learn how to perform quantum chemical calculations with them. However, it is harder to learn to recognize problems that are amenable to computation, choose methods appropriate to the problem, and assess the results. The latter tasks are difficult because many choices have to be made. Several books have been written on the subject.

This chapter summarizes a minimal amount of essential material about quantum chemical methods to help you understand papers that use such methods. An introduction to basis set selection is given in section 10.2. The choice of functional in a KS-DFT calculation, the handling of heavy atoms, and methods to account for solvent effects are discussed very briefly in section 10.2. An outline of practical computational strategies and a map of the main methods of current computational chemistry are given in section 10.3.

doi:10.1088/978-1-627-05416-4ch10

10.2 So many choices to be made

10.2.1 Selection of a basis set

In the H_2^+ calculations of section 7.1 and section 7.4, the MOs were represented by linear combinations of atom-centered $1s$-type functions ($e^{-\zeta r}$) and $2p_z$-like polarization functions ($ze^{-\zeta r}$). These functions are special cases of *Slater-type functions* defined by $\mathcal{N} e^{-\zeta r} r^\ell \mathcal{Y}_{\ell,m}$ where $\mathcal{Y}_{\ell,m}$ is a spherical harmonic (section 5.2) and $\mathcal{N}$ is a normalization constant; see table 10.1.

Unfortunately, the computation of the required integrals is rather time-consuming for Slater-type functions. The integrals are greatly simplified if one uses Gaussian functions defined by $\mathcal{N} e^{-\zeta r^2} r^\ell \mathcal{Y}_{\ell,m}$ and listed in table 10.1. However, this leads to a loss of accuracy. A solution that combines computational efficiency and accuracy is to use fixed linear combinations of Gaussians, called *contracted Gaussian functions* (CGF), with parameters predetermined by atomic calculations or some other prescription. For example, $s_1 = 0.317\, e^{-0.110\, r^2} + 0.381\, e^{-0.406\, r^2} + 0.109\, e^{-2.228\, r^2}$ is an s-type CGF. Almost all contemporary calculations use a basis set of CGFs.

A *minimal basis set* contains one basis function for each type of AO that is occupied in the ground state of each atom in the molecule. For example, a minimal basis set contains one s-type basis function for each hydrogen atom in a molecule since H has one occupied AO ($1s$). Similarly, a minimal basis set has two s-type and one set of three p-type basis functions for each second-period atom since these atoms have an occupied $1s$ AO in the core, and occupied $2s$ and $2p$ valence AOs[1]. Thus, a minimal basis set for benzene (C_6H_6) contains $N = 6 \times 5 + 6 \times 1 = 36$ basis functions. A well-known minimal CGF basis set is STO-3G in which the name denotes that CGF consisting of three Gaussians are used for each AO.

Minimal basis sets are not accurate enough for most purposes and are not used except for very large systems. Instead, minimal basis sets are conceptual building blocks for *double-zeta* (DZ) and *split-valence* (SV) basis sets. A double-zeta basis contains twice as many functions as a minimal basis; each function in the minimal basis set is replaced by two of the same type. We have already seen a double-zeta basis set for He on page 8-7. The description of an atom *in a molecule* may require

Table 10.1. Slater- and Gaussian-type functions in unnormalized form.

Type	Slater	Gaussian	Type	Slater	Gaussian
s	$e^{-\zeta r}$	$e^{-\zeta r^2}$	d_{xy}	$xy\, e^{-\zeta r}$	$xy\, e^{-\zeta r^2}$
p_x	$x\, e^{-\zeta r}$	$x\, e^{-\zeta r^2}$	d_{yz}	$yz\, e^{-\zeta r}$	$yz\, e^{-\zeta r^2}$
p_y	$y\, e^{-\zeta r}$	$y\, e^{-\zeta r^2}$	d_{xz}	$xz\, e^{-\zeta r}$	$xz\, e^{-\zeta r^2}$
p_z	$z\, e^{-\zeta r}$	$z\, e^{-\zeta r^2}$	$d_{x^2-y^2}$	$(x^2 - y^2)\, e^{-\zeta r}$	$(x^2 - y^2)\, e^{-\zeta r^2}$
			d_{z^2}	$(3z^2 - r^2)\, e^{-\zeta r}$	$(3z^2 - r^2)\, e^{-\zeta r^2}$

---

[1] It is necessary to add p-type basis functions in Li and Be because they have unoccupied $2p$-type AOs that are quite close in energy to the occupied $2s$ AO.

p functions of different sizes in different directions because the chemical environment of the atom may be different in different directions.

This cannot be achieved with a minimal basis set but is clearly possible with a DZ basis set. A split-valence basis set is a minimal basis set for the core and a DZ set for the valence region. A SV set is almost as accurate as a DZ set for many chemical properties because they are not affected much by the core electrons. For example, a SV basis set for a second-period atom such as C contains nine CGF: three of *s* type and six of *p* type. Common SV basis sets include the 3-21G, 4-31G, and 6-31G basis sets[2]. The notation 3-21G tells us that each core CGF contains three Gaussians and there are two valence CGF—one containing two Gaussians and the other containing only one Gaussian.

The smallest basis sets that offer a good compromise between accuracy and computational effort are obtained by adding *polarization functions* of higher angular momentum to a DZ or SV basis set. The polarization functions help to describe how the electron cloud of an atom polarizes (distorts) under the influence of the other atoms in the molecule. The polarization of an *s* function by a p_z function was illustrated on page 7-7. The polarization of *p*-type functions is achieved with *d*-type functions, as shown below.

An SV plus polarization (SVP) basis set is an SV basis plus a set of three *p*-type functions on each hydrogen atom, a set of five *d*-type functions on each B, C, N, O, and F atom, and so on. A popular SVP basis set is the 6-31G** or 6-31G(d,p) set, which consists of a 6-31G set plus polarization functions. Another common SVP basis set is cc-pVDZ. Sometimes polarization functions are included only on the non-hydrogen atoms, as in the 6-31G* or 6-31G(d) basis set, to reduce computation.

Calculations on anions and weakly bonded systems require *diffuse functions*—Gaussians with relatively small exponents. The addition of diffuse functions to a basis set is often denoted by ++ or the 'aug-' prefix. For example, the 6-31++G(d,p) basis set consists of the 6-31G(d,p) basis plus an *s*-type diffuse function on each H atom, an

---

[2] These basis sets and others named in this fashion were developed in the laboratory of John Pople who shared the 1998 Nobel Prize in Chemistry 'for his development of computational methods in quantum chemistry'. The widely used GAUSSIAN computer program was developed by his research group.

s-type and a set of *p*-type diffuse Gaussians on each B, C, N, O, and F atom, and so on. The cc-pVDZ set with added diffuse functions is called aug-cc-pVDZ.

Larger basis sets are constructed in a similar manner. For example, one can have valence triple-zeta basis sets with two sets of polarization functions, as in the 6-311G (2d,2p) basis set. A systematic sequence of basis sets of increasing quality is formed by the cc-pVDZ, cc-pVTZ, cc-pVQZ, cc-pV5Z, and cc-pV6Z sets, and another sequence is formed by their aug- counterparts. The computer time required to do matrix HF or KS calculations increases with the number of basis functions N as N^4. For large molecules, the computational cost increases less rapidly, roughly as $N^{2.7}$ for HF and KS-DFT, because of the weak coupling between basis functions that are located on well-separated atoms.

10.2.2 Selecting a functional

An exchange-correlation potential v_{xc}, or an exchange-correlation energy functional[3] E_{xc}, has to be specified if KS-DFT calculations are desired. Many functionals are available but it is probably best to stick to the most heavily used and well-tested ones. Functionals of the generalized gradient approximation (GGA) type depend upon both the electron density and its gradient. Popular GGA functionals include the empirical BLYP and SOGGA functionals and the theoretically better justified PW91 and PBE (sometimes called PBEPBE) functionals. Computationally more expensive hybrid methods use a mixture of the HF exchange operator $\hat{K}$ and an approximate v_{xc}. B3LYP is by far the most widely used hybrid functional although PBE0 (sometimes called PBE1PBE) is better justified theoretically. Corrections for van der Waals (dispersion) forces are now common as in the B3LYP-D3 scheme. Finding better functionals is a high priority in current research; new functionals are proposed almost every month.

10.2.3 Heavy atoms and relativistic effects

The observation that many chemical properties of atoms and molecules are determined primarily by the valence electrons was exploited in the construction of smaller, but equally accurate, basis sets on page 10-2. For molecules containing atoms from the fifth period and beyond, that is Rb and heavier atoms, a much greater computational saving can be obtained if all terms describing the interaction of the electrons in the core orbitals with each other, with the valence electrons, and with the nuclei, are simply replaced by an '*effective core potential*' (ECP), and the nuclear charges Z appearing in the interaction potential between the valence electrons and the nuclei are replaced by effective nuclear charges $Z_{eff} = Z - N_c$, where N_c is the number of core electrons. In addition to the computational savings, ECPs can be made to approximate relativistic effects as well; the latter are important in atoms with a large nuclear charge because the innermost electrons move with a speed approaching that of light. All this has to be done in a manner that does not

---

[3] The two are related by $v_{xc}(\vec{r}) = \delta E_{xc}/\delta\rho(\vec{r})$.

violate the Pauli principle. The process of constructing an ECP is complex and well beyond the scope of this book.

The choice of N_c is important but not always intuitive. For example, to obtain good results, it suffices to treat Si ([Ne] $3s^2 3p^2$) as a 4 valence-electron system but it is necessary to treat Ti as a 12 valence-electron system ([Ne] $3s^2 3p^6 3d^2 4s^2$) and not as a 4 valence-electron system ([Ar] $3d^2 4s^2$). ECPs are often available for varying core sizes. It is usually safe to use small-core or medium-core ECPs but large-core ECPs should be used only with caution. The most commonly used effective core potentials include those from the Los Alamos National Laboratory and the Stuttgart–Dresden collections.

10.2.4 Accounting for a solvent

All the methods discussed so far relate to molecules in the gas phase but the properties of a molecule in solution can be quite different from its gas-phase properties. Reaction-field methods are currently used to include solvent effects. The solvent is modeled as a continuum of uniform dielectric constant ε, and the solute molecule is placed in a cavity within the solvent field. The simple Onsager model, in which a spherical cavity is used, predicts a non-zero solvent effect only for molecules with a non-zero dipole moment. The polarizable continuum model is a more sophisticated model in which the cavity is described by a set of interlocking atomic spheres. Reaction-field calculations are routine; the only additional actions required from the user are to specify the solvent, choose the reaction-field model, and run a calculation to estimate the cavity radius if the Onsager model is to be used.

10.3 Practical calculations

Quantum chemical calculations almost never lead to exact results; we just find inexact solutions to approximate equations. Hence, *it rarely makes sense to do just a single calculation.* Instead, a series of calculations must be done to assess the reliability of the results. The size of the molecules and degree of difficulty in computing the property of interest will determine tactical details, such as the choice of basis sets and methods. Often, the limitations of the available computer hardware force one to settle for a lower level of calculation than one knows, or finds, is required to obtain the desired reliability.

The goal of the calculations should determine the computational strategy. If the object is to understand or predict trends in the variation of a property across a set of molecules, then one should calculate the property using an inexpensive KS-DFT method which lies in the middle third of the simplified map of quantum chemical methods shown in figure 10.1. It is advisable to calibrate the results by performing or looking up parallel calculations on an additional set of similar molecules for which the trend is already known. A KS-DFT computation requires a choice of both basis set and functional. The usual notation for a KS-DFT calculation is the name of the functional followed by a slash followed by the basis set name. For example, one can talk about a B3LYP/6-31G(d) or PBE/cc-pVTZ calculation. Hybrid functionals are often more accurate than GGA varieties, especially for

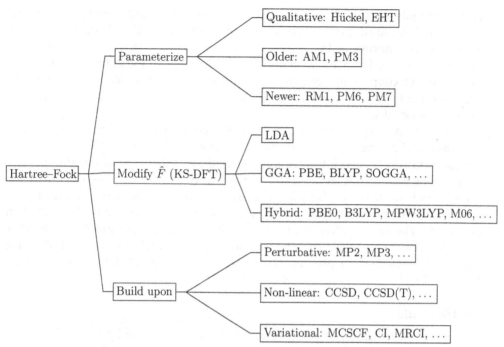

Figure 10.1. A schematic map of the quantum chemical forest.

organic molecules. The most commonly reported calculations are of the B3LYP/ 6-31G(d,p) and B3LYP/6-31+G(d,p) varieties. However, it is more prudent to use PBE or PBE0 for molecules containing several transition metal atoms. A correction for dispersion (van der Waals) interactions is often applied to KS-DFT calculations. It is not uncommon to see KS-DFT calculations in a systematically improving sequence of basis sets.

There are many other *ab initio*[4] methods for finding wave functions better than the HF model but they are beyond the scope of this book. The simplest such method, called second-order Møller–Plesset (MP2) perturbation theory, is widely used to calculate a correction to the HF energy in terms of integrals that involve all the occupied *and* virtual HF orbitals. In situations where KS-DFT is known to produce results of uneven quality, it is recommended to do MP2 calculations for calibration. A similar notation is used for MP2 computations—for example, one writes MP2/cc-pVTZ or MP2/6-311G(2d,2p). MP2 calculations typically take only twice as long as KS-DFT calculations. If both KS-DFT and MP2 results turn out to be not accurate enough for one's purposes, then one must do systematic *ab initio* calculations as described in the next paragraph.

If the object is to calculate as accurate a value as possible for some property of a small molecule, then it is appropriate to perform a sequence of computations in which the level of both the method and the basis set are systematically increased

---

[4] *Ab initio* derives from Latin and means 'from first principles'.

until the results stabilize. KS-DFT calculations cannot be improved in a systematic manner and one must turn to methods which begin with a HF calculation and then systematically improve it; the latter methods are shown in the bottom third of the map in figure 10.1. The MP2 method is usually the first method in this sequence, and the CCSD(T) coupled-cluster method is often the last. Calculations of this sort require some technical knowledge and experience, and so consulting an expert is usually a good idea.

If the system being studied is very large, KS-DFT calculations may be too time-consuming. Then one can use *semiempirical* methods based on simplified matrix HF equations in which most integrals are neglected and others are assigned numerical values using empirical rules or data. Semiempirical methods are 'take-it-or-leave-it' since there is no basis set, functional, or method to vary. A simple semiempirical method, useful only for qualitative purposes, is the Hückel model detailed in chapter 9. The most widely used, quantitative, semiempirical methods are AM1 (Austin model 1) and PM3 (parametric model 3)[5]. Recently, AM1 and PM3 have been reparameterized to produce the improved versions RM1 (Recife model 1), PM6, and PM7 (parametric models 6 and 7).

Further study

A practical introduction to computational chemistry is given by James Foresman and Æleen Frisch in their *Exploring Chemistry with Electronic Structure Methods* (Gaussian, Inc., Pittsburgh, 2nd edition, 1996). The **GAMESS** computer program, and its Fortran source code, are freely available from http://www.msg.ameslab.gov/gamess/ and can be used to try out calculations. Use Ira Levine's *Quantum Chemistry*, (Prentice Hall, NJ, 5th edition, 2000) to learn much more about the methods introduced in this book and others. Case studies and numerical comparisons of the performance of various basis sets, functionals, and methods can be found in Christopher Cramer's *Essentials of Computational Chemistry: Theories and Models*, (Wiley, New York, 2nd edition, 2004).

---

[5] Do not confuse PM3 with MP3 (3rd-order Møller–Plesset perturbation theory).

Quantum Chemistry
A concise introduction for students of physics, chemistry, biochemistry and materials science
Ajit J Thakkar

Appendix A

Reference material

Matrices and determinants

An $N \times M$ matrix A is a rectangular array of numbers with N rows and M columns; the number in the ith row and jth column is denoted A_{ij}. An $N \times 1$ matrix is a column vector with N components. Most matrices used in quantum chemistry are either square matrices or column vectors. The *transpose* of an $N \times M$ matrix A is a $M \times N$ matrix denoted A^T which has the elements $A_{ij}^T = A_{ji}$. A square matrix is called *symmetric* if it is equal to its transpose. The transpose of a column vector is a row vector.

Matrices of the same size can be added element by element; $C = A + B$ means that $C_{ij} = A_{ij} + B_{ij}$ for all i, j. If A is an $N \times N$ matrix and c is a scalar, then cA is an $N \times N$ matrix with elements cA_{ij}. Matrices of commensurate sizes can be multiplied. If A is an $N \times M$ matrix and B is a $M \times P$ matrix, then $C = AB$ is an $N \times P$ matrix whose ijth element is given by the dot product of the ith row of A and jth column of B. Thus,

$$C_{ij} = A_{i1}B_{1j} + A_{i2}B_{2j} + \cdots + A_{iM}B_{Mj}. \tag{A.1}$$

The product of two square $N \times N$ matrices is an $N \times N$ square matrix. Multiplication of square matrices is not necessarily commutative; AB can be and sometimes is different from BA. A square $N \times N$ matrix times an $N \times 1$ column vector is an $N \times 1$ column vector. A square matrix I with ones on the diagonal and zeroes elsewhere is called an *identity matrix*. When A and I are square $N \times N$ matrices and C is an $N \times 1$ column vector, $IA = AI = A$ and $IC = C$.

A square matrix has a number, called its *determinant*, associated with it. For a 1×1 matrix A, $\det A = A_{11}$. For a 2×2 matrix A,

$$\det A = \det \begin{vmatrix} A_{11} & A_{12} \\ A_{21} & A_{22} \end{vmatrix} = A_{11}A_{22} - A_{21}A_{12}. \tag{A.2}$$

doi:10.1088/978-1-627-05416-4ch11

The determinant of an $N \times N$ matrix can be expanded 'by minors' as a sum of N determinants of $(N-1) \times (N-1)$ matrices. For example, the determinant of a 3×3 matrix is given by a sum of three 2×2 determinants:

$$\det \begin{vmatrix} A_{11} & A_{12} & A_{13} \\ A_{21} & A_{22} & A_{23} \\ A_{31} & A_{32} & A_{33} \end{vmatrix}$$

$$= A_{11} \det \begin{vmatrix} A_{22} & A_{23} \\ A_{32} & A_{33} \end{vmatrix} - A_{12} \det \begin{vmatrix} A_{21} & A_{23} \\ A_{31} & A_{33} \end{vmatrix} + A_{13} \det \begin{vmatrix} A_{21} & A_{22} \\ A_{31} & A_{32} \end{vmatrix}. \tag{A.3}$$

Each scalar above is from the first row of the 3×3 determinant with alternating signs, and the jth 2×2 determinant is obtained by striking out the first row and jth column in the parent 3×3 determinant. Interchanging a pair of rows or columns of a determinant changes its sign. Hence, if a pair of rows or columns of A is identical, then $\det A = 0$.

A real, symmetric, square $N \times N$ matrix A has N eigenvalues λ_i and eigen(column)vectors C_i that satisfy the eigenvalue equation:

$$AC_i = \lambda_i C_i. \tag{A.4}$$

The trace of a square $N \times N$ matrix A is the sum of its diagonal elements: $\operatorname{Tr} A = A_{11} + A_{22} + \cdots + A_{NN}$. The sum of the eigenvalues of A equals its trace: $\lambda_1 + \lambda_2 + \cdots + \lambda_N = \operatorname{Tr} A$.

Miscellaneous

Complex numbers can be expressed as $z = a + ib$ where a and b, respectively, are called the real and imaginary parts of z, and $i = \sqrt{-1}$ is called the imaginary unit. The complex conjugate of z is denoted z^* and is given by $z^* = a - ib$. Generally, setting i to $-i$ wherever it appears suffices to find the complex conjugate. The squared magnitude of z is given by

$$|z|^2 = z^* z = (a - ib)(a + ib) = a^2 - iba + aib - i^2 b^2 = a^2 + b^2. \tag{A.5}$$

A *polynomial* $P(x)$ of degree n has the form

$$P(x) = a_0 + a_1 x + a_2 x^2 + \cdots + a_n x^n \tag{A.6}$$

where the a_i are constants and are called coefficients. $P(x)$ has n roots—that is, there are n values of x for which $P(x) = 0$.

The *factorial* function is defined by

$$n! = n(n-1)(n-2)\ldots 1 \tag{A.7}$$

and $0! = 1$. For example, $1! = 1$, $2! = 2$, $3! = 6$, and $4! = 24$.

Table of integrals

$$\int x \sin ax \, dx = \frac{1}{a^2} \sin ax - \frac{x}{a} \cos ax \tag{A.8}$$

$$\int \sin^2 ax \, dx = \frac{x}{2} - \frac{1}{4a} \sin 2ax \tag{A.9}$$

$$\int x \sin^2 ax \, dx = \frac{x^2}{4} - \frac{x}{4a} \sin 2ax - \frac{1}{8a^2} \cos 2ax \tag{A.10}$$

$$\int x^2 \sin^2 ax \, dx = \frac{x^3}{6} - \left(\frac{x^2}{4a} - \frac{1}{8a^3}\right) \sin 2ax - \frac{x}{4a^2} \cos 2ax \tag{A.11}$$

$$\int x^2 e^{ax} \, dx = \frac{e^{ax}}{a} \left(x^2 - \frac{2x}{a} + \frac{2}{a^2}\right) \tag{A.12}$$

$$\int_a^b f'(x)g(x) \, dx = [f(x)g(x)]_a^b - \int_a^b f(x)g'(x) \, dx \tag{A.13}$$

$$\int_0^\infty x^n e^{-ax} \, dx = \frac{n!}{a^{n+1}} \qquad \text{for } a > 0 \text{ and } n = 0, 1, 2, \ldots \tag{A.14}$$

$$\int_{-\infty}^\infty e^{-ax^2} \, dx = \left(\frac{\pi}{a}\right)^{1/2} \qquad \text{for } a > 0 \tag{A.15}$$

$$\int_{-\infty}^\infty x^{2n} e^{-ax^2} \, dx = \frac{1 \cdot 3 \cdot 5 \cdots (2n-1)}{2^n a^n} \left(\frac{\pi}{a}\right)^{1/2} \qquad \text{for } a > 0 \text{ and } n = 1, 2, \ldots \tag{A.16}$$

$$\int_{-\infty}^\infty x^{2n+1} e^{-ax^2} \, dx = 0 \qquad \text{for } a > 0 \text{ and } n = 0, 1, 2, \ldots \tag{A.17}$$

$$\int_0^\infty x^{2n+1} e^{-ax^2} \, dx = \frac{n!}{2a^{n+1}} \qquad \text{for } a > 0 \text{ and } n = 0, 1, 2, \ldots \tag{A.18}$$

Conversion factors

$1\,a_0$	$= 52.918$ pm
	$= 0.529\,18$ Å
$1\,\text{Å}$	$= 1.8897\,a_0$
	$= 100$ pm
$1\,E_h$	$= 2625.5$ kJ mol^{-1}
	$= 627.51$ kcal mol^{-1}
	$= 27.211$ eV
	$= 2.1947 \times 10^5$ cm^{-1}
	$= 4.3597 \times 10^{-18}$ J
$1\,\text{kJ mol}^{-1}$	$= 3.8088 \times 10^{-4}\,E_h$
$1\,\text{kcal mol}^{-1}$	$= 1.5936 \times 10^{-3}\,E_h$
$1\,\text{eV}$	$= 3.6749 \times 10^{-2}\,E_h$
$1\,\text{cm}^{-1}$	$= 4.5563 \times 10^{-6}\,E_h$
$1\,u$	$= 1.6605 \times 10^{-27}$ kg

Prefix		Meaning
peta	P	10^{15}
tera	T	10^{12}
giga	G	10^9
mega	M	10^6
kilo	k	$10^3 = 1000$
deci	d	$10^{-1} = 0.1$
centi	c	$10^{-2} = 0.01$
milli	m	$10^{-3} = 0.001$
micro	μ	10^{-6}
nano	n	10^{-9}
pico	p	10^{-12}
femto	f	10^{-15}

Constants and Greek letters

	Value
c	2.9979×10^8 m s^{-1}
h	6.6261×10^{-34} J s
$\hbar$	1.0546×10^{-34} J s
e	1.6022×10^{-19} C
m_e	9.1094×10^{-31} kg
N_A	6.0221×10^{23} mol^{-1}

From: Mohr P J, Taylor B N and Newell D B
2012 *Rev. Mod. Phys.* **84** 1527

α	alpha		ν	nu
β	beta		Ξ, ξ	xi
Γ, γ	gamma		o	omicron
Δ, δ	delta		Π, π	pi
ϵ, ε	epsilon		ρ, ϱ	rho
ζ	zeta		Σ, σ	sigma
η	eta		τ	tau
$\Theta, \theta, \vartheta$	theta		Υ, υ	upsilon
ι	iota		Φ, ϕ, φ	phi
κ	kappa		χ	chi
Λ, λ	lambda		Ψ, ψ	psi
μ	mu		Ω, ω	omega

Equation list

Photon properties $\qquad E = h\nu \quad \text{and} \quad \lambda\nu = c$

Schrödinger equation $\qquad \hat{H}\psi_n = E_n\psi_n$

Orthonormality $\qquad \int \psi_m^2 \, d\tau = 1 \quad \text{and} \quad \int \psi_m\psi_n \, d\tau = 0 \quad \text{for } m \neq n$

Quantum operators $\qquad \hat{x} = x, \quad \hat{p}_x = -i\hbar\partial/\partial x, \quad \hat{p}_x^2 = -\hbar^2\partial^2/\partial x^2$

$\qquad \hat{T}_x = \hat{p}_x^2/(2m), \quad \hat{H} = \hat{T} + \hat{V}$

Mean value $\qquad \langle A \rangle = \int \psi(\hat{A}\psi) \, d\tau$

Variance $\qquad \sigma(A) = (\langle A^2 \rangle - \langle A \rangle^2)^{1/2}$

Heisenberg uncertainty principle $\qquad \sigma(x)\sigma(p_x) \geqslant \hbar/2$

Particle in a wire $\qquad E_n = n^2h^2/(8ma^2) \quad \text{for } n = 1, 2, \ldots$

$\qquad \psi_n(x) = (2/a)^{1/2} \sin(n\pi x/a) \quad \text{for } 0 \leqslant x \leqslant a$

Particle in a $\qquad E_{n_x,n_y} = h^2(n_x^2/a^2 + n_y^2/b^2)/(8m)$

rectangular plate $\qquad \psi_{n_x,n_y}(x,y) = [4/(ab)]^{1/2} \sin(n_x\pi x/a) \sin(n_y\pi y/b)$

$\qquad \text{inside the box, for } n_x, n_y = 1, 2, \ldots$

Harmonic oscillator $\qquad E_v = (v + 1/2)\hbar\omega \quad \text{for } v = 0, 1, 2, \ldots; \omega = \sqrt{k/m}$

$\qquad \psi_0(x) = (\alpha/\pi)^{1/4} e^{-\alpha x^2/2} \quad \text{with } \alpha = m\omega/\hbar$

Particle on a ring $\qquad E_m = \hbar^2 m^2/(2I) \quad \text{for } m = 0, \pm 1, \pm 2, \ldots$

$\qquad \psi_0(\phi) = \sqrt{1/(2\pi)}$

Particle on a sphere $\qquad E_{\ell,m} = \hbar^2\ell(\ell+1)/(2I) \quad \text{for } \ell = 0, 1, \ldots; \ell \geqslant |m|$

$\qquad \psi_{0,0}(\theta,\phi) = \mathcal{Y}_{0,0}(\theta,\phi) = \sqrt{1/(4\pi)}$

$\qquad \langle \hat{L}^2 \rangle = \ell(\ell+1)\hbar^2, \quad \langle \hat{L}_z \rangle = m\hbar$

Rigid rotor $\qquad I = \mu R^2 \quad \text{with } \mu = m_1 m_2/(m_1 + m_2)$

Cartesian/spherical $\qquad r = (x^2 + y^2 + z^2)^{1/2}$

$\qquad dx \, dy \, dz = r^2 \sin\theta \, dr \, d\theta \, d\phi$

Hydrogen atom $\qquad E_{n,\ell,m} = -E_h/(2n^2) \quad \text{for } n = 1, 2, \ldots; n > \ell \geqslant |m|$

$\qquad \psi_{1,0,0}(\vec{r}) = 1s(r) = \pi^{-1/2} e^{-r}$

Hydrogen-like ion $\qquad E_{n,\ell,m} = -Z^2 E_h/2n^2 \quad \text{for } n = 1, 2, \ldots; n > \ell \geqslant |m|$

$\qquad \psi_{1,0,0}(\vec{r}) = 1s(r) = (Z^3/\pi)^{1/2} e^{-Zr}$

Variational principle $\qquad \int \Phi(\hat{H}\Phi) \, d\tau / \int |\Phi|^2 \, d\tau \geqslant E_{gs}$

Hartree–Fock $\qquad \hat{F}\varphi_k = \epsilon_k\varphi_k \quad \text{for } k = 1, 2, \ldots$

Matrix Hartree–Fock $\qquad \mathbf{FC}_k = \epsilon_k\mathbf{SC}_k \quad \text{for } k = 1, 2, \ldots$

Hartree–Fock electron density $\qquad \rho(\vec{r}) = m_1|\varphi_1(\vec{r})|^2 + m_2|\varphi_2(\vec{r})|^2 + \cdots + m_n|\varphi_n(\vec{r})|^2$

Hückel matrix $\qquad \mathbf{F} = \alpha\mathbf{I} + \beta\mathbf{A}$

Hückel MO (HMO) $\qquad \varphi_k = c_{1k}p_1 + c_{2k}p_2 + \cdots + c_{Nk}p_N$

HMO normalization $\qquad c_{1k}^2 + c_{2k}^2 + \cdots + c_{Nk}^2 = 1$

Hückel π-charges $\qquad q_j = m_1|c_{j1}|^2 + m_2|c_{j2}|^2 + \cdots + m_N|c_{jN}|^2$

Hückel bond orders $\qquad P_{ij} = m_1 c_{i1}c_{j1} + m_2 c_{i2}c_{j2} + \cdots + m_N c_{iN}c_{jN}$

Hückel free valence $\qquad f_k = \sqrt{3} - B_k$

IOP Concise Physics

Quantum Chemistry
A concise introduction for students of physics, chemistry, biochemistry and materials science
Ajit J Thakkar

Appendix B

Problem hints and solutions

1.1 (a) SF_6, (b) trans-HFC=CHF, (c) $H_2C=CH_2$, (d) Benzene, (e) Staggered ethane, and (f) $(CHBrCl)_2$.

1.2 Hint: Apply the C_4 rotation to unit vectors along the x and y axes.

1.3 Naphthalene has three C_2 axes, one σ_h and two σ_v planes of symmetry, and a center of symmetry at the center of mass. The principal C_2 axis is along the CC bond shared by the rings. Another C_2 axis is perpendicular to the molecular plane and passes through the center of mass. The third C_2 axis is perpendicular to the other two and bisects the bond shared by the rings. If the origin is at the center of mass, and the x, y, and z axes lie along the three C_2 axes, then the xy, yz, and xz planes are symmetry planes.

1.4 Both molecules are planar. The molecular plane is a σ for the molecule on the left and a σ_h for the molecule on the right, which also has a C_2 axis and a center of symmetry.

1.5 Each of the symmetry planes contains all three carbon atoms and two hydrogen atoms. The principal C_2 axis passes through the three carbon atoms. The other two C_2 axes can be seen with effort and help from the Newman diagram below. The line of sight is the principal C_2 axis through all carbons. The unbroken solid line represents the front hydrogen atoms and the broken line the back hydrogen atoms. The dashed lines are the C_2 axes; they are perpendicular to the principal axis and pass through the central carbon.

1.6 A table is the easiest way to present the results:

Molecule	Point group	Polar?	Chiral?
Pyridine	C_{2v}	Yes	No
Borazine	D_{3h}	No	No
CH_3F	C_{3v}	Yes	No
$SnCl_4$	T_d	No	No

1.7 (a) C_{2v}, (b) D_{3h}, (c) C_s, and (d) D_{2d}.

1.8 A dipole moment vector can lie along at most one of the three C_2 axes, and would be moved by the other two axes. Therefore, the molecule cannot be polar. It has no planes of symmetry, no inversion center, and no improper axis of symmetry. Therefore, it could be chiral. See section 1.3.

2.1 (a) $a^2 e^{-ax}$, (b) $a^4/4 - (2/3)a^3 + (3/2)a^2 - 4a$, and (c) $12x^2y^3z^2 + 6x^4yz^2 + 2x^4y^3$.

2.2 (a) $\hat{A}^2 = x^2$. (b) $\hat{A}^2 = d^2/dx^2$. (c) Recall from equation (2.2) that x and d/dx do not commute.

$$\hat{A}^2 f = (d/dx + x)^2 f = (d/dx + x)(df/dx + xf)$$
$$= d^2f/dx^2 + d(xf)/dx + x(df/dx) + x^2 f$$
$$= d^2f/dx^2 + (dx/dx)f + x(df/dx) + x(df/dx) + x^2 f$$
$$= d^2f/dx^2 + f + 2x(df/dx) + x^2 f$$

and hence

$$\hat{A}^2 = d^2/dx^2 + 2x(d/dx) + (x^2 + 1).$$

2.3 The results are best presented in a table:

Function	eigenfunction of $-\hbar^2 d^2/dx^2$?	eigenvalue
e^{-ax^2}	No	—
$\cos \beta x$	Yes	$\hbar^2 \beta^2$
$7e^{ikx}$	Yes	$\hbar^2 k^2$

2.4 Hint: Sketch the functions over the interval $-1 \leqslant x \leqslant 1$. $f(x)$ is square integrable but $g(x)$ is not. Changing the value of the constant does not matter as long as it is kept positive.

2.5 The wave functions differ only by a factor of -1, a phase factor. They lead to exactly the same probability density ψ^2.

2.6 (a) $f(x)$ is normalized:

$$\int_0^\infty f(x) f(x) \, dx = 1,$$

(b) $g(x)$ is normalized:

$$\int_0^\infty g(x)\,g(x)\,\mathrm{d}x = 1,$$

and (c) $f(x)$ is orthogonal to $g(x)$:

$$\int_0^\infty g(x)\,f(x)\,\mathrm{d}x = \int_0^\infty f(x)\,g(x)\,\mathrm{d}x = 0.$$

2.7 Using equation (2.16), the integral (A.14) from appendix A with $n = 0$, and choosing the positive root leads to $c = \sqrt{2a}$.

3.1 It would double.

3.2 The wavelength $\lambda = c/\nu$ where c is the speed of light and ν is the frequency of the photon. By Einstein's relationship, $E = h\nu$ and hence $\lambda = hc/E$. The energy E is the difference between the energies of the $n = 2$ and $n = 1$ levels, and is therefore $E = h^2(2^2 - 1^2)/(8ma^2)$ in which m is the mass of the electron (m_e) and a is the length of the wire. Finally, we have $\lambda = 8cm_e a^2/(3h)$. Inserting the various quantities in SI units leads to $\lambda = 275$ nm which is in the near ultraviolet (UV).

3.3 Figure 3.3 shows that $n = 4$ is one of many such states. Think about equation (3.5). Which other states satisfy these requirements?

3.4 Postulate 1 on page 2-2 tells us that calculation of the probabilities will require integrals of $|\psi_1(x)|^2$, the square of the ground-state wave function. Equation (3.5) gives us $\psi_1^2 = (2/a)\sin^2(\pi x/a)$. The integrals can be done using equation (A.9) from appendix A or with mathematical software. (a) The probability of finding the particle in the left half of the wire is

$$\int_0^{a/2} |\psi_1(x)|^2\,\mathrm{d}x = \frac{2}{a}\int_0^{a/2} \sin^2(\pi x/a)\,\mathrm{d}x = 1/2.$$

(b) The probability of finding the particle in a quarter at the edge is

$$\int_0^{a/4} |\psi_1(x)|^2\,\mathrm{d}x = \int_{3a/4}^{a} |\psi_1(x)|^2\,\mathrm{d}x = 0.091,$$

and for a middle quarter it is

$$\int_{a/4}^{a/2} |\psi_1(x)|^2\,\mathrm{d}x = \int_{a/2}^{3a/4} |\psi_1(x)|^2\,\mathrm{d}x = 0.409.$$

Observe that the probabilities for the four quarters add up to 1.

3.5 Section 3.1 contains a calculation of $\langle x\rangle$. Use equation (2.17) with the ground-state wave function from equation (3.5) and the integral formula of equation (A.11) from appendix A to find that

$$\langle x^2\rangle = \int_0^a |\psi_1(x)|^2\,x^2\,\mathrm{d}x = \frac{(2\pi^2 - 3)}{6\pi^2}a^2 = 0.283\,a^2.$$

Use $\hat{p}_x = -i\hbar\, \partial/\partial x$ and the identity $2\sin y \cos y = \sin(2y)$ to find that

$$\langle p_x \rangle = -i\hbar \int_0^a \psi_1(x)(\mathrm{d}\psi_1/\mathrm{d}x)\,\mathrm{d}x = -i\hbar\frac{2}{a}\int_0^a \sin(\pi x/a)\frac{\mathrm{d}\,\sin(\pi x/a)}{\mathrm{d}x}\,\mathrm{d}x$$

$$= -i\hbar\frac{2}{a}\frac{\pi}{a}\int_0^a \sin(\pi x/a)\cos(\pi x/a)\,\mathrm{d}x = -i\hbar\frac{\pi}{a^2}\int_0^a \sin(2\pi x/a)\,\mathrm{d}x = 0.$$

Note that $\hat{p}_x^2 = -\hbar^2\,\partial^2/\partial x^2$ and do some calculation to find:

$$\langle p_x^2 \rangle = -\hbar^2\int_0^a \psi_1(x)(\mathrm{d}^2\psi_1/\mathrm{d}x^2)\,\mathrm{d}x = h^2/(4a^2).$$

The average value of the kinetic energy is $\langle \hat{T}_x \rangle = \langle \hat{p}_x^2 \rangle/(2m)$. Substituting the value calculated above for $\langle \hat{p}_x^2 \rangle$, we find that $\langle \hat{T}_x \rangle = h^2/(8ma^2)$ which equals the ground-state energy given by equation (3.6) with $n = 1$. This makes sense because all the energy is kinetic in this problem. Finally,

$$\sigma(x)\sigma(p_x) = \frac{\sqrt{3\pi^2 - 18}}{6}\hbar = 0.568\,\hbar,$$

which clearly satisfies the Heisenberg inequality of equation (2.19).

3.6

(a) Assuming that the four π electrons occupy the two lowest states, the lowest energy transition is the one from the highest occupied level ($n = 2$) to the lowest unoccupied one ($n = 3$). Using the difference between the energies of the $n = 3$ and $n = 2$ levels of a particle in a wire, $\lambda\nu = c$, Einstein's relation $E = h\nu$, and the mass of the electron m_e, you should find $\lambda = 8cm_eL^2/(5h)$ in which the length of the wire could be taken as $L = 2R(C{=}C) + R(C{-}C)$. This estimate leads to a length of $L \approx 2\times134 + 154 = 422$ pm for the backbone of 1,3-butadiene, and $\lambda = 117$ nm. Other reasonable estimates of the length of the wire would be acceptable.

(b) $\lambda = 8cm_eL^2/(7h)$, $L \approx 710$ pm, and $\lambda = 237$ nm.

(c) Hints: What will be the quantum number of the highest occupied energy level? How many $C{=}C$ and $C{-}C$ bonds are there?

3.7 It would double because $\omega = (k/m)^{1/2}$.

3.8 Hint: See page 3-7 for a derivation of one of the expectation values needed. The ground state of a harmonic oscillator is called a minimum uncertainty state.

3.9 Hints: Do not forget to use the reduced mass of the diatomic. The wave numbers need to be converted to frequencies and all quantities should be in SI units before final use. $k = 515.57$ N m^{-1} for $^1\text{H}^{35}\text{Cl}$.

4.1 Populating the energy levels shown in figure 4.3 with four π electrons in accordance with the Pauli principle and Hund's rule, we find the scheme shown below. Clearly, cyclobutadiene is predicted to be a *diradical* and hence so very reactive that it is nearly impossible to isolate.

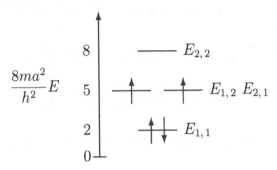

4.2 Hint: Once you understand how figure 4.4 follows from equation (4.9), then you will be able to use equation (4.9) to make the sketches.

4.3 The first and third terms in the Hamiltonian define a one-dimensional harmonic oscillator in the x direction and the second and fourth terms define a one-dimensional harmonic oscillator in the y direction. Thus, the separation of variables technique (see page 4-1) allows us to write:

$$E_{v_x,v_y} = \hbar\omega_x(v_x + 1/2) + \hbar\omega_y(v_y + 1/2)$$

in which $\omega_x = (k_x/m)^{1/2}$ and $\omega_y = (k_y/m)^{1/2}$; two quantum numbers, $v_x = 0, 1, \ldots$ and $v_y = 0, 1, \ldots$, are needed in this two-dimensional problem.

If $k_y = k_x$, then $\omega_y = \omega_x$ and $E_{v_x,v_y} = \hbar\omega_x(v_x + v_y + 1)$. Pairs of states with interchanged quantum numbers, such as (1,2) and (2,1), become degenerate. Some accidental degeneracies also arise exactly as in the case of a particle in a square plate; see section 4.2.

4.4 In this three-dimensional problem, there are three quantum numbers, $n_x = 1, 2, \ldots, n_y = 1, 2, \ldots$, and $n_z = 1, 2, \ldots$. The wave functions outside the box are simply $\psi_{n_x,n_y,n_z}(x, y, z) = 0$. Inside the box, the three-dimensional wave functions are given by

$$\psi_{n_x,n_y,n_z}(x, y, z) = \left(\frac{8}{abc}\right)^{1/2} \sin\left(\frac{n_x\pi x}{a}\right) \sin\left(\frac{n_y\pi y}{b}\right) \sin\left(\frac{n_z\pi z}{c}\right).$$

The energies are given by

$$E_{n_x,n_y,n_z} = \frac{h^2}{8m} \left(\frac{n_x^2}{a^2} + \frac{n_y^2}{b^2} + \frac{n_z^2}{c^2}\right).$$

This is the *fundamental* model of translational motion in three dimensions. For example, it is used in statistical thermodynamics to calculate the properties of an ideal monatomic gas, and the translational contributions to the properties of an ideal molecular gas.

4.5 Setting $b = a$ and $c = a$ in the solutions to problem 4.4, we find that the energies for a particle in a cube are given by

$$E_{n_x,n_y,n_z} = \frac{h^2}{8ma^2}\left(n_x^2 + n_y^2 + n_z^2\right).$$

The three states with $(n_x, n_y, n_z) = (2, 1, 1)$, $(1, 2, 1)$, and $(1, 1, 2)$ all have the same energy $3h^2/(4ma^2)$. The first excited energy level is three-fold degenerate.

4.6 Borazine, CH_3F and $SnCl_4$ will have degeneracies in their energy levels because of the presence of a C_3 axis. See section 4.2.

4.7 Cyclopropane (D_{3h}) and spiropentane (D_{2d}) have degeneracies in their energy levels. See table 4.1.

4.8 There is an even number of electrons. If there are no degeneracies in the energy levels, as is the case when a molecule belongs to the D_{2h} point group (see table 4.1), opposite-spin pairs of electrons will occupy the lowest levels and there will be no unpaired electrons. On the other hand, in the case of a molecule with D_{6h} symmetry, Hund's rule may lead to unpaired electrons in the highest occupied energy level(s) because the latter could be degenerate.

4.9 Replace x by $(a - x)$ in equation (3.5) to get

$$\psi_n(a - x) = \sqrt{2/a}\,\sin[n\pi(a - x)/a].$$

Note that the trigonometric identity $\sin(\alpha - \beta) = \sin\alpha\cos\beta - \sin\beta\cos\alpha$ can be applied to $\psi_n(a - x)$ above with $\alpha = n\pi a/a = n\pi$ and $\beta = n\pi x/a$. This gives us

$$\psi_n(a - x) = \sqrt{2/a}\,[\sin(n\pi)\cos(n\pi x/a) - \sin(n\pi x/a)\cos(n\pi)].$$

Substitute the values $\sin(n\pi) = 0$, $\cos(n\pi) = +1$ for $n = 2, 4, \ldots$, and $\cos(n\pi) = -1$ for $n = 1, 3, \ldots$, in the above to find, as required, that

$$\psi_n(a - x) = \hat{i}\psi_n(x) = \begin{cases} +\psi_n(x) & \text{for } n = 1, 3, 5, \ldots \\ -\psi_n(x) & \text{for } n = 2, 4, 6, \ldots \end{cases}$$

5.1 There are two important qualitative differences:
1. There are degeneracies in the energy levels of a particle on a circular ring but not in the energy levels of a particle in a wire.
2. The ground-state energy of a particle on a circular ring is zero but the ground state energy of a particle in a wire is non-zero. In other words, there is no rotational zero-point energy but there is a translational zero-point energy.

5.2 The spacing would decrease by a factor of four because R^2 appears in the denominator of the energy formula, equation (5.3).

5.3 Verify normalization by showing that

$$\int_0^{2\pi} |\psi_1|^2\,d\phi = \int_0^{2\pi} |\psi_{-1}|^2\,d\phi = 1$$

and orthogonality by showing that

$$\int_0^{2\pi} \psi_1\psi_{-1}\, d\phi = 0.$$

Use either mathematical software or integral formulas to do the integrals.

5.4 0 because of orthogonality.

5.5 The reduced mass of HBr is

$$\mu = m_H m_{Br}/(m_H + m_{Br}) = 1.008 \times 79.9/(1.008 + 79.9) = 0.99544\ u.$$

Using a conversion factor from appendix A, $\mu = 0.99544 \times 1.6605 \times 10^{-27} = 1.6529 \times 10^{-27}$ kg. Using equation (5.9), the difference between the energies of the $\ell = 0$ and $\ell = 1$ levels is

$$\Delta E_{0\to1} = E_{1,m} - E_{0,m} = \frac{\hbar^2}{2I}[1(1+1) - 0(0+1)] = \frac{\hbar^2}{I} = \frac{\hbar^2}{\mu R^2}.$$

The frequency of a photon that can cause this transition is $\nu = \Delta E_{0\to1}/h$; insert the SI values of h, $\hbar$, μ in kg, and $R = 1.414 \times 10^{-10}$ m to find $\nu = 5.079 \times 10^{11}$ Hz $= 507.9$ GHz. Similarly, the frequency for the $\ell = 1 \to 2$ transition is 1016 GHz.

5.6 Hint: Follow the example shown in section 5.3 to find $R = 113.1$ pm.

5.7 Hint: This is similar to problem 3.6, for which a solution has been given, except that the energy levels of a particle on a circular ring have to be used instead of those for a particle in a straight wire.

5.8 (a) $1/(4\pi)$, (b) $1/(4\pi)$, (c) Unsöld's theorem states that

$$|\mathcal{Y}_{\ell,-\ell}|^2 + |\mathcal{Y}_{\ell,-\ell+1}|^2 + \cdots + |\mathcal{Y}_{\ell,\ell-1}|^2 + |\mathcal{Y}_{\ell,\ell}|^2 = \frac{2\ell+1}{4\pi}.$$

6.1 The vibrational transition energy is $E_1 - E_0 = \hbar\omega$. The transition energy is proportional to the frequency and to the wave number, which is the reciprocal of the wavelength. Recall that $\omega = \sqrt{k/\mu}$. The isotopologues H_2 and D_2 will have the same k but different μ. Thus we can write

$$\frac{\omega(D_2)}{\omega(H_2)} = \frac{\sqrt{k/\mu(D_2)}}{\sqrt{k/\mu(H_2)}} = \sqrt{\frac{\mu(H_2)}{\mu(D_2)}}.$$

From http://physics.nist.gov/cgi-bin/Compositions/stand_alone.pl we find the atomic masses of H and D to be 1.0078 and 2.0141, respectively. Since $\mu = m_1 m_2/(m_1 + m_2)$, we have $\mu(H_2) = 0.50390$ and $\mu(D_2) = 1.00705$ and

$$\omega(D_2) = 4395.2 \times \sqrt{\frac{0.50390}{1.00705}} = 3109.0\ \text{cm}^{-1}.$$

The heavier molecule has a smaller vibrational frequency as expected.

6.2 For $n = 5$, the allowed values of ℓ are $0, 1, 2, 3, 4$ or s, p, d, f, g. For f orbitals, $\ell = 3$ and the allowed values of m are $-3, -2, -1, 0, 1, 2, 3$.

6.3 (a) 2, (b) 8, and (c) 18. Why are these degeneracies equal to the number of atoms in the (a) 1st, (b) 2nd, and (c) 4th periods of the periodic table?

6.4 Use equation (6.7) to find that

$$\Delta E = E_{2,1,0} - E_{1,0,0} = \frac{-1}{2 \times 2^2} - \frac{-1}{2 \times 1^2} = \frac{3}{8} E_h.$$

Combining $\Delta E = h\nu$ and $\lambda\nu = c$, we have $\lambda = hc/\Delta E$. Keeping everything in SI units (ΔE has to be converted from E_h to J), we find $\lambda = 121.5$ nm.

6.5 The energy of the $2p_z$ state ($n = 2$) is $-(1/8)E_h$ and the ionization threshold is at $E = 0$. So the ionization energy is $(1/8)E_h = 3.401$ eV.

6.6 $\langle r \rangle = 5\,a_0$ for the $2p_z$ state just as it does for the $2p_x$ state because the value of $\langle r \rangle$ depends only on the n and ℓ quantum numbers as shown by equations (6.11)–(6.13).

6.7 Using table 6.1, equation (6.13), and either integral formulas from appendix A or mathematical software, we find $\langle r \rangle = (25/2)\,a_0$ and $(21/2)\,a_0$ for the $3p$ and $3d$ states respectively.

6.8 Substitute $R_{1s}(r) = 4\sqrt{2}\,e^{-2r}$ from page 6-9 into equation (6.11) to find $D(r) = 32r^2\,e^{-4r}$. Inserting $D(r)$ into equation (6.13) and using equation (A.14) leads to

$$\langle r \rangle = 32 \int_0^\infty r^3\,e^{-4r}\,dr = 32 \times 3!/4^4 = (3/4)\,a_0.$$

Note that $\langle r \rangle = (3/4)\,a_0$ for He^+ is half the size of $\langle r \rangle = (3/2)\,a_0$ for H because the electron density of He^+ is more compact.

6.9 Taking R_{2s} for the H atom from table 6.1 and following the prescription in section 6.4, we find for He^+ ($Z = 2$) that

$$R_{2s}(r) = \left(\sqrt{2^3}/\sqrt{2}\right)(1 - 2r/2)\,e^{-2r/2} = 2(1 - r)\,e^{-r}.$$

The 2s orbital for He^+ is $2s = R_{2s}\mathcal{Y}_{0,0} = R_{2s}/\sqrt{4\pi} = \pi^{-1/2}(1 - r)\,e^{-r}$.

6.10 The integral equals zero because the orbitals (wave functions) of the hydrogen atom are orthogonal to each other.

6.11
(a) Using equation (6.11) with $R_{1s}(r) = 2\,e^{-r}$, one obtains

$$P(\mathcal{R}) = 1 - e^{-2\mathcal{R}}\left(1 + 2\mathcal{R} + 2\mathcal{R}^2\right).$$

(b) $P(\mathcal{R}) = 0.85, 0.95$, and 0.99 when $\mathcal{R} = 2.36\,a_0$, $3.15\,a_0$, and $4.20\,a_0$ respectively.

7.1 The electronic Hamiltonian for the HeH^{2+} molecule with bond length R is given by

$$\hat{H}_e = -\frac{1}{2}\nabla_1^2 - \frac{2}{r_{1a}} - \frac{1}{r_{1b}} + \frac{2}{R}$$

where r_{1a} is the distance between the electron and the He nucleus (labeled a) and r_{1b} is the distance between the electron and the proton (labeled b). The first term corresponds to $\hat{T}_e$, the next two to V_{en}, and the last one to V_{nn}. $V_{ee} = 0$ because there is only one electron. Compare with the Hamiltonian for H$_2^+$ on page 7-1.

7.2 Integrate the product of the $1\sigma_g$ and $1\sigma_u$ MOs of equation (7.4) as follows:

$$\int 1\sigma_g 1\sigma_u \, d\tau = N_g N_u \int \left(1s_a^2 - 1s_a 1s_b + 1s_b 1s_a - 1s_b^2\right) d\tau$$

$$= N_g N_u \left(\int 1s_a^2 \, d\tau - \int 1s_b^2 \, d\tau\right) = N_g N_u(1-1) = 0.$$

7.3 H$_2$ has two electrons and its configuration is $1\sigma_g^2$. He$_2$ has four electrons and its configuration is $1\sigma_g^2 1\sigma_u^2$. He$_2$ is not a stable molecule because the bonding effect of $1\sigma_g^2$ is canceled out by the antibonding effect of $1\sigma_u^2$.

7.4 $\zeta = 1$ and $E = -(1/2)E_h$. His trial function contains the exact wave function as a special case and so the variational method should find it.

7.5 The energies become $E_g = H_{aa} + H_{ab}$ and $E_u = H_{aa} - H_{ab}$. The two energies are symmetrically placed below and above H_{aa}, the non-bonding energy. When overlap is included, E_u is further above H_{aa} than E_g is below H_{aa}; in other words, $1\sigma_u$ is more antibonding than $1\sigma_g$ is bonding.

7.6

(a) $\int \Phi(\hat{H}\Phi) \, dx = \dfrac{\hbar^2 a^3}{6m} = \dfrac{h^2 a^3}{24\pi^2 m}$.

(b) $\int |\Phi|^2 \, dx = a^5/30$.

(c) $E_\Phi = \dfrac{5}{4\pi^2} \dfrac{h^2}{ma^2} = 0.12665 \dfrac{h^2}{ma^2}$. This is 1.3% higher than the exact ground-state energy $\dfrac{h^2}{8ma^2}$.

7.7 $\pi_{g,xz} = d_{xz,a} + d_{xz,b}$ and $\pi_{u,xz} = d_{xz,a} - d_{xz,b}$. The MOs are labeled π because the d_{xz} AOs have $m = 1$; see the convention on page 7-2 and table 5.1. The g and u designations indicate that the MOs are symmetric and antisymmetric with respect to inversion. Make sketches of the MOs and observe that $\pi_{g,xz}$ is antibonding and $\pi_{u,xz}$ is bonding. The $\pi_{g,xz}$ MO and the analogous $\pi_{g,yz}$ MO form a doubly degenerate pair. Similarly, the $\pi_{u,xz}$ and $\pi_{u,yz}$ MOs form a doubly degenerate pair.

7.8 $\delta_{g,xy} = d_{xy,a} + d_{xy,b}$ and $\delta_{u,xy} = d_{xy,a} - d_{xy,b}$. The MOs are labeled δ because the d_{xy} AOs have $m = 2$. The g and u designations indicate that the MOs are symmetric and antisymmetric with respect to inversion. $\delta_{g,xy}$ is bonding and $\delta_{u,xy}$ is antibonding. The $(\delta_{g,xy}, \delta_{g,x^2-y^2})$ MOs are doubly degenerate as are the $(\delta_{u,xy}, \delta_{u,x^2-y^2})$ MOs.

7.9

(a) We find, in appendix A, values of E_h in both kJ mol^{-1} and cm^{-1} which we can use to convert ω to kJ mol^{-1} as follows:

$$\omega = 2297 \text{ cm}^{-1} \times 2625.5 \text{ kJ mol}^{-1}/(2.1947 \times 10^5 \text{ cm}^{-1}) = 27.48 \text{ kJ mol}^{-1}.$$

Thus $D_0 = D_e - \omega/2 = 242$ kJ mol^{-1} (see section 7.5) for $^1\text{H}_2^+$.

(b) Since D_2^+ and H_2^+ are isotopologues, they have the same electronic Hamiltonian. Hence, within the Born–Oppenheimer approximation, they have the same potential energy curve; see section 6.1. Therefore, D_2^+ has $R_e = 106$ pm just as H_2^+ does.

(c) For the reasons explained above, $D_e = 269$ kJ mol^{-1} for D_2^+.

(d) Following the solution to problem 6.1, find $\omega = 1625$ cm^{-1} for D_2^+.

(e) $D_0 = D_e - \omega/2 = 250$ kJ mol^{-1} for D_2^+. D_0 is larger for D_2^+ than for H_2^+ because the heavier D_2^+ has a smaller vibrational zero-point energy.

8.1 The Li$^+$ cation consists of two electrons and a nucleus of charge $+3$. Hence,

$$\hat{H}_e = -\frac{1}{2}\nabla_1^2 - \frac{1}{2}\nabla_2^2 - \frac{3}{r_1} - \frac{3}{r_2} + \frac{1}{r_{12}}$$

where r_1 and r_2, respectively, are the distances between electrons 1 and 2 and the nucleus, and r_{12} is the interelectronic distance. The first two terms constitute $\hat{T}_e$, the next two terms make up V_{en}, and the last term is V_{ee}. $V_{nn} = 0$ because there is only one nucleus. Compare with $\hat{H}_e$ for He.

8.2 The given wave function does not satisfy the Pauli postulate. It is symmetric rather than antisymmetric with respect to interchange of all the coordinates of the electrons.

8.3 Observe that $|1s\overline{1s}1s| = 0$ because a determinant vanishes when two of its rows are identical. We cannot have a wave function that is zero everywhere because the electron probability density cannot be zero everywhere.

8.4

(a) Using $1s\beta$ and $2s\beta$ as the two occupied spin orbitals, we find

$$\psi(1,2) = \frac{1}{\sqrt{2}}\det \begin{vmatrix} \overline{1s}(1) & \overline{1s}(2) \\ \overline{2s}(1) & \overline{2s}(2) \end{vmatrix}.$$

(b) The expanded form is a product of spatial and spin functions:

$$\psi(\vec{r}_1, \sigma_1, \vec{r}_2, \sigma_2) = [1s(\vec{r}_1)2s(\vec{r}_2) - 2s(\vec{r}_1)1s(\vec{r}_2)]\beta(\sigma_1)\beta(\sigma_2)/\sqrt{2}.$$

8.5

(a) ROHF wave function: $|1s\overline{1s}2s\overline{2s}2p_z|$,

(b) UHF wave function: $|1s\overline{1s'}2s\overline{2s'}2p_z|$.

8.6 Koopmans' approximation predicts that the (first) ionization energy is approximately $-\epsilon_{HOMO}$. CH_4 has 10 electrons and 5 doubly occupied MOs. In CH_4, the HOMO is any one of the triply degenerate $1t_2$ MOs. Hence, from the given data, the (first) ionization energy of CH_4 is $0.54\ E_h$.

8.7

(a) Use the variational method. Choose ζ to minimize the energy. Set $dE/d\zeta = 2\zeta - 2Z + 5/8 = 0$ and find $\zeta = Z - 5/16$. For the neutral helium atom with $Z = 2$, this gives $\zeta = 2 - 5/16 = 27/16$ as in section 8.1.

(b) Insert the above $\zeta = Z - 5/16$ into the given energy expression, and do a bit of algebraic manipulation to find $E = -(Z - 5/16)^2$.

(c) Setting $Z = 1$ in the above formula gives $E = -(11/16)^2 \approx -0.47\ E_h$, which is above the exact energy of a neutral H atom ($-0.5\ E_h$). This calculation predicts that the hydride anion H^- can gain energy by losing an electron. This is an incorrect result and another failure of the orbital model due to its neglect of dynamical electron correlation.

8.8 The electron configuration of Ar is $1s^2 2s^2 2p^6 3s^2 3p^6$. There should be five peaks in the photoelectron spectrum of Ar corresponding, in order of increasing energy, to ejection of an electron from a $3p$, $3s$, $2p$, $2s$, and $1s$ orbital. The first four peaks would be seen in a vacuum UV photoelectron spectrum and the highest-energy peak in an x-ray photoelectron spectrum. The usual (first) ionization energy is the energy corresponding to ejection of an electron from the HOMO ($3p$).

8.9 O_2 has the electron configuration $1\sigma_g^2 1\sigma_u^2 2\sigma_g^2 2\sigma_u^2 3\sigma_g^2 1\pi_{ux}^2 1\pi_{uy}^2 1\pi_{gx}^1 1\pi_{gy}^1$. Removing an electron from one of the doubly degenerate HOMOs gives the electron configuration of O_2^+ as either $1\sigma_g^2 1\sigma_u^2 2\sigma_g^2 2\sigma_u^2 3\sigma_g^2 1\pi_{ux}^2 1\pi_{uy}^2 1\pi_{gx}^1$ or $1\sigma_g^2 1\sigma_u^2 2\sigma_g^2 2\sigma_u^2 3\sigma_g^2 1\pi_{ux}^2 1\pi_{uy}^2 1\pi_{gy}^1$. Since the HOMOs are both antibonding orbitals, removal of an electron from either one of them results in strengthening the bond and shortening it. Thus, O_2 should have a longer bond length than O_2^+.

8.10 The ground-state electron configuration of Li_2 is $1\sigma_g^2 1\sigma_u^2 2\sigma_g^2$. Base your orbital energy diagram on the bottom of figure 8.4. Sketches of the occupied MOs can be made as in figure 7.6. There are two electrons in each of the bonding σ_g MOs and two electrons in the antibonding σ_u MO. Remembering to count electron pairs as opposed to electrons, the nominal bond order is $2 - 1 = 1$. Li_2^+ would have one less electron in the bonding HOMO and a nominal bond order of $1.5 - 1 = 0.5$. Hence, the Li_2^+ cation is less strongly bound than its neutral parent.

8.11 The ground-state electron configuration of Be_2 is $1\sigma_g^2 1\sigma_u^2 2\sigma_g^2 2\sigma_u^2$. Base your orbital energy diagram on the bottom of figure 8.4. Sketches of the occupied MOs can be made as in figure 7.6. There are two electrons in each of the bonding σ_g MOs and two electrons in each of the antibonding σ_u MOs. Remembering to count electron pairs as opposed to electrons, the nominal bond order is $2 - 2 = 0$. The LUMO is $1\pi_u$.